3rd International Conference on Smart Material Research (ICSMR 2017)

IOP Conference Series: Materials Science and Engineering
Volume 293

Melbourne, Australia
15 – 17 November 2017

ISBN: 978-1-5108-5534-2
ISSN: 1757-8981

Printed from e-media with permission by:

Curran Associates, Inc.
57 Morehouse Lane
Red Hook, NY 12571

Some format issues inherent in the e-media version may also appear in this print version.

Printed by Curran Associates, Inc. (2018)

For permission requests, please contact the Institute of Physics
at the address below.

Institute of Physics
Dirac House, Temple Back
Bristol BS1 6BE UK

Phone: 44 1 17 929 7481
Fax: 44 1 17 920 0979

techtracking@iop.org

Additional copies of this publication are available from:

Curran Associates, Inc.
57 Morehouse Lane
Red Hook, NY 12571 USA
Phone: 845-758-0400
Fax: 845-758-2633
Email: curran@proceedings.com
Web: www.proceedings.com

Table of contents

Volume 293

3rd International Conference on Smart Material Research (ICSMR 2017)
15–17 November 2017, Melbourne, Australia

Accepted papers received: 21 December 2017
Published online: 5 January 2018

Preface

Papers

Biomedical Materials and Bioelectronics

PREFACE

It is with great pleasure that I introduce the proceedings of the 2017 3rd International Conference on Smart Material Research (ICSMR 2017) which was held recently in Melbourne, Australia, 15-17, November 2017. ICSMR 2017 was dedicated to technical issues related to smart material research.

The objectives of the conference were not only to provide opportunities to the international scientists to present state of the art research in smart materials but also to provide a forum to discuss ideas for future directions in the field and explore possibilities for collaborative research projects.

The conference program included keynote, oral, and poster presentations from scholars working in the areas of materials science and engineering from all over the world. It covered recent trends and progress made in the field of materials engineering and manufacturing. Professors from Australia, USA, Italy, and Belgium were invited to deliver keynote lectures regarding the latest information in their respective areas of expertise.

The proceedings present a selection of high quality papers submitted to the conference by researchers from universities, research institutes, and industry. All papers were subjected to peer-review by conference committee members and international reviewers. The papers were selected based on their quality and their relevance to the conference. The proceedings present recent advances in the fields of materials science and engineering and manufacturing as well as in other related areas including materials properties, measuring methods, analytical and numerical modeling and simulations among others.

I would like to express special gratitude to members of the conference committee and organizers of the conference. I would also like to thank the reviewers for their valuable time and advice which helped in improving the quality of the papers selected for presentation at the conference and for publication in the proceedings. Finally I want to thank the authors, the members of the organizing committee, the reviewers, the chairpersons, sponsors, and all other conference participants for their support of ICSMR 2017.

Prof. Ramesh K. Agarwal
Washington University in St. Louis, USA
November 22, 2017

IOP Conf. Series: Materials Science and Engineering **293** (2017) 011001 doi:10.1088/1757-899X/293/1/011001

STATEMENT OF PEER REVIEW

All papers published in this volume of *IOP Conference Series: Materials Science and Engineering* have been peer reviewed through processes administered by the proceedings Editors. Reviews were conducted by expert referees to the professional and scientific standards expected of a proceedings journal published by IOP Publishing.

IOP Conf. Series: Materials Science and Engineering **293** (2017) 011002 doi:10.1088/1757-899X/293/1/011002

Conference Committee Chairs

Darren Martin, University of Queensland, Australia
Ramesh K. Agarwal, Washington University in St. Louis, USA

Conference Local Chair

Warren Batchelor, Monash University, Australia

Program Committee Chairs

Maya Kiskinova, Elettra-Sincrotrone, Italy
Takuya Tsuzuki, The Australian National University, Australia
Rongkun Zheng, The University of Sydney, Australia
Christine Jérôme, Université de Liège, CERM, Belgium

International Technical Committees

Alexander Pesin,Nosov Magnitogorsk State Technical University, Russia
Leslie Joy Diaz, University of the Philippines, Philippines
Ramin Yousefi, Islamic Azad University, Iran
Carlos Rolando Rios-Soberanis, Centro de Investigación Científica de Yucatán, México
Irshad Ahmad, American University of Ras Al Khaimah, UAE
Siti Rahmah Aid, Malaysia-Japan International Institute of Technology(MJIIT),Universiti Teknologi Malaysia, Malaysia
Yuri Vorobiev, CINVESTAV-IPN, Unidad Queretaro, México
AGUS ARSAD, Universiti Teknologi Malaysia, Malaysia
Heba Al-Kelesh, Central Metallurgical Research and Development Institute, Egypt
Khairulzan Yahya, Universiti Teknologi Malaysia
Redzuan Abdullah, Universiti Teknologi Malaysia
Jamaludin Mohd. Yatim, Universiti Teknologi Malaysia
Izni Syahrizal Ibrahim, Universiti Teknologi Malaysia
Mohammadreza Vafaei, Universiti Teknologi Malaysia
MOHD YUHAZRI BIN YAAKOB, Universiti Teknikal Malaysia Melaka, Malaysia

IOP Conf. Series: Materials Science and Engineering **293** (2017) 011003 doi:10.1088/1757-899X/293/1/011003

Peer review statement

All papers published in this volume of *IOP Conference Series: Materials Science and Engineering* have been peer reviewed through processes administered by the proceedings Editors. Reviews were conducted by expert referees to the professional and scientific standards expected of a proceedings journal published by IOP Publishing.

IOP Conf. Series: Materials Science and Engineering **293** (2017) 012001 doi:10.1088/1757-899X/293/1/012001

Effect of temperature, viscosity and surface tension on gelatine structures produced by modified 3D printer

C Kalkandelen[1], B Ozbek[2,3], N M Ergul[2,3], S Akyol[4], Y Moukbil[2,5] , F N Oktar[2,5], N Ekren[2,6] O Kılıc[2,7], B Kılıc[8], O Gunduz[2,9]

[1] Department of Biomed. Devices Tech., Vocational School of Technical Sciences, Istanbul Univ., Istanbul, Turkey, 34320

[2] Advanced Nanomater. Research Lab., Faculty of Technology, Marmara Univ., Istanbul, Turkey, 34722

[3] Department of Metallurgical and Materials Engineer., Institute of Pure and Applied Sciences, Marmara Univ., Istanbul, Turkey, 34722

[4] Istanbul University, Cerrahpasa Medical School, Istanbul, Turkey, 34096

[5] Department of Bioengineer., Faculty of Engineer., Marmara Univ., Istanbul, Turkey, 34722

[6] Department of Electric and Electronic Eng., Faculty of Technology, Marmara Univ., Istanbul, Turkey, 34722

[7] Department of Electric and Electronic Eng., Faculty of Engineer., Marmara Univ., Istanbul, Turkey,34722

[8] Yıldız Technical University, Faculty of Electric-Electronic Engineering, Department of Electric Eng., 34220, Istanbul / TURKEY

[9] Department of Metallurg. and Material. Engineer., Faculty of Technology, Marmara Univ., Istanbul, Turkey, 34722

ucemogu@ucl.ac.uk

Abstract. In the present study, gelatine scaffolds were manufactured by using modified 3D (3 Dimensional) printing machine and the effect of different parameters on scaffold structure were investigated. Such as; temperature, viscosity and surface tension of the gelatine solutions. The varying of gelatine solutions (1, 3, 5, 10, 15 and 20 wt.%) were prepared and characterized. It has been detected that, viscosity of those solutions were highly influenced by temperature and gelatine concentration. Specific CAD (Computer Assistant Design) model which has 67% porosity and original design were created via computer software. However, at high temperatures gelatine solutions caused like liquid but at the lower temperatures were observed the opposite behaviour. In addition to that, viscosity of 1,3,5 wt.% solutions were not enough to build a structure and 20 wt.% gelatine solution too hard to handle, because of the sudden viscosity changes with temperature. Even though, scaffold of the 20 wt.% gelatine solution printed hardly but it was observed the best printed solutions, which were 10 and 15 wt.% gelatine solutions. As a result, 3D printing of gelatine were found the values of the best temperature, viscosity, surface tension and gelatine concentration such as 25-35 °C, 36-163 cP, 46-59 mN/m and 15 wt.% gelatine concentration respectively.

1. Introduction

It was observed that the life quality of the humans has increased with the researches about tissue engineering and regenerative medicine, which aims producing biocompatible structures mimicking biological organs, tissues or systems. Scaffolds, which were manufactured for this purpose, provides cell like matrix and they need to meet some mechanical and design features, which contributes functionality of the structures [1]. Consequently, various materials and methods have been used to obtain the printing scaffolds. Lately, 3D printing technology came forward cause of ability of producing scaffolds designed similar to the structures found in the human body. This technology allows controlling the aimed scaffold by layer-by-layer. Also porosity, pore size and pore morphology are adjustable by designing CAD (Computer Assistant Design) models with some computer software [2]. With the ability of to control of design of scaffolds and fabrication of complex shapes, this technology became crucial in tissue engineering. Some of the 3D printing applications in tissue engineering are renovation of anatomic defects, rebuilding of organs with 3D microarchitecture and scaffolds for stem cell differentiation [3].

Gelatine is a natural material and this water soluble protein obtained from collagen by denaturation of it. Due to its biocompatibility and biodegradability, gelatine has come to an important place for the regenerative medicine applications [1]. Also it contains biological mark, which promotes cell adhesion and proliferation [2]. As a result of that, for the researches working in tissue engineering, this material, which has biological origin, derived from collagen, has become an appealing choice [4]. But aqueous solutions of the gelatine form a hydrogel structure, which behaves like solid, when it gets below their critical solution temperature of 25-35°C [5]. On the contrary, too much heat application can cause them to behave like liquid. For these reasons, temperature of the solution should be kept in a stable state. This thermo depended behaviour can be used to build layers on one while retaining the shape of printed structures [6].

In this study, we focused to investigate the behaviour of the gelatine solution and printability of it by changing its viscosity, surface tension and density. These changes influenced by temperature differences and gelatine concentration of the solutions. Viscosity and surface tension of the solutions were measured at temperature range between 25-55°C. Following by that gelatine scaffolds with specific design printed with modified 3D printer. As a result, very suitable printing conditions for the scaffolds were obtained.

2. Materials and methods

Aqueous solutions of the gelatine were prepared by adding 1, 3, 5, 10, 15 and 20 wt.% gelatine to distilled water. Solutions were stirred with magnetic stirrer with heat application until gelatine fully dissolved. After dissolving, viscosity and surface tension were performed between the temperature range of 25-55°C. Viscosity measurement of the solutions were fulfilled via viscometer (Brookfield, DV-E, USA). Desired temperature achieved by circulating water inside the machine. Density of the solutions were obtained by a standard density bottle DIN ISO 3507-Gay-Lussac (Boru Cam, 10 ml., Turkey). For surface tension measurements digital densitometer (Sigma, 703D, Finland) was used and the Du Nouy ring method was conducted. Surface tension of the water used for solutions preparation was 71mN/m.

Prepared solutions loaded into 10 ml syringe and polyethylene tube was used to for streaming of solution. To obtain the desired temperature, tube was wrapped with resistance and voltage applied to resistance to get heat. Before mounting syringe to syringe pump, one side of the tube connected to syringe, while other side of the tube connected to print head. At the tip of the printer head, tube connected with stainless steel needle via luer lock adaptor. When connection completed, syringe mounted into syringe pump (Inovenso, IPS-12, Turkey) and solution was purged for printing. After printing, surfaces of the samples were investigated by using optical microscope (Olympus, BX51M, Japan).

Ultimaker 2+ (Ultimaker, Netherlands) was used for this study, but for the purpose of the present study, original print head of the printer was dismantled from the machine and a new print head was designed and printed with another Ultimaker 2+ (Figure 1). Print speed parameter changed over the device, while flow rate and temperature parameters changed from the outside. For this study, 8-10 ml/h flow rate used and print speed changed between 20-50%. Design of the sample was obtained via Solidworks CAD software. Values of dimensions of layers of the design selected as 30 x 20 x 0.7 mm (Figure 2). Samples designed to have 67% porosity. The reason of the using samples, which have high porosity rate, is that high porosity increase the potential of the application of the scaffolds clinically [1].

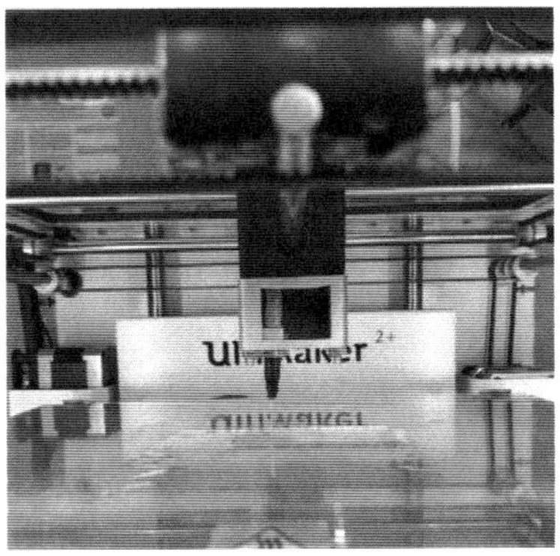

Figure 1. Modified print head, luer lock and stainless steel needle.

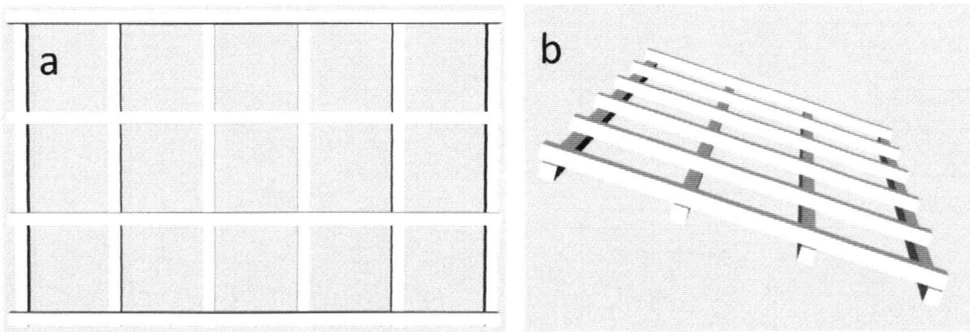

Figure 2. CAD design of the samples. a) top view of the design b) side view of the design.

3. Results and discussion

To investigate the effect of solution density, all of the solutions were putted in density bottle. As a result, it was observed that density change with a slight difference (Table 1). Increasing gelatine concentration influenced the viscosity and the surface tension more than the density.

Table 1. Density of the solutions.

Gelatine (wt.%)	Density (g/cm^3)
1	1.028 ± 0.021
3	1.031 ± 0.025
5	1.039 ± 0.023
10	1.045 ± 0.031
15	1.051 ± 0.034
20	1.0518 ± 0.035

As mentioned earlier, it has been observed that the temperature change and gelatine concentration greatly influence the viscosity. In this study, the viscosity of the solutions are obtained in temparature range of 20-55 °C, for the solutions viscosity has been found as follows, for 1 wt.% gelatine 0.88-1.85, for 3 wt.% 2.20-20.00 cP, for 5 wt.% 3.75-27 cP, for 10 wt.% 9.78-120 cP, for 15 wt.% 25-430 cP and for 20 wt.% 56-480 cP (Figure 3). This behaviour is harmanious with Arrhenius Law, which suggests that logarithmic decrease of viscosity with increasing temperature. Schuurman *et al.* suggest that, increasing the viscosity of the solution will affect the connection of pores and ability to print wider pores. As a result it will be much easier to print scaffolds with larger pores [1]. In parallel with that, scaffolds produced with gelatine solutions have higher viscosity revealed more stifness than the others. But 20 wt.% gelatine solution at room temperature blocked the needle. So while 10 wt.% and 15 wt.% gelatine could printed between 25-35°C, 20 wt.% gelatine need more temperature.

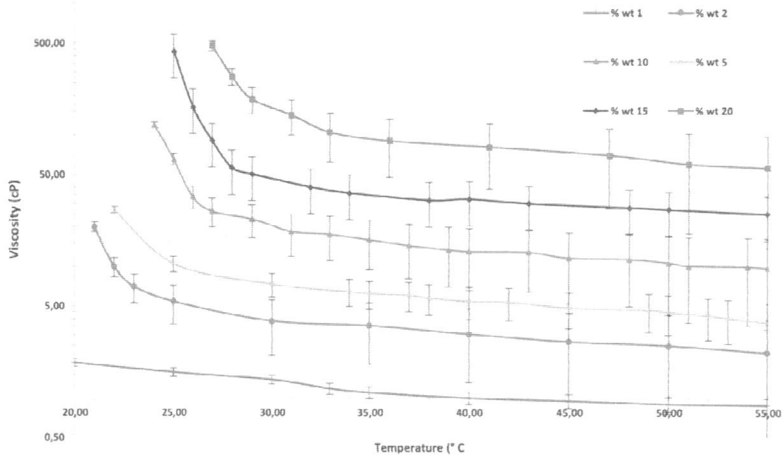

Figure 3. Viscosity of the solutions with logarithmic y-axis.

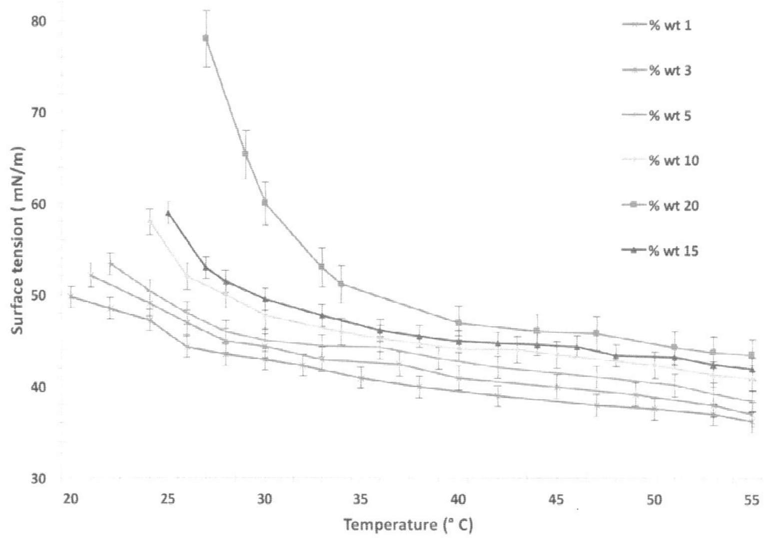

Figure 4. Surface tension of the solutions.

Similar to viscosity, surface tension of the solutions also has been affected with temperature and gelatine concentration. No study was found in the literature related to surface tension and 3D printing of gelatine structures. It is clearly seen that the surface tension is directly proportional according to the viscosity and the surface tension is inversely proportional according to the temperature. Surface tension of the solutions have been found as follows; for 1 wt.% 36.2-49.77 mN/m, for 3 wt.% 37.0-52.13 mN/m, for 5 wt.% 38.5-53.4 mN/m, for 10 wt.% 41.0-58.0 mN/m, for 15 wt.% 42.0-59 mN/m and for 20 wt.% 43.5-78 mN/m (Figure 4). It has been shown that, higher surface tension cause difficulty to building second layer on first layer. Therefore, desired stifness and the rigid structure of the samples can not be printed.

As a result, optimal gelatine concetration for printing was found as 10 and 15 wt.% with the working temperature at 25-35°C . The 20 wt.% gelatin solution was found with printable as 35-45°C. Investigation of the printed samples were obtained via the optical microscope (Figure 5). Images show that, the 15 and 20 wt.% gelatine have much more smooth and uniform surface than 10 wt.% of gelatine sample.

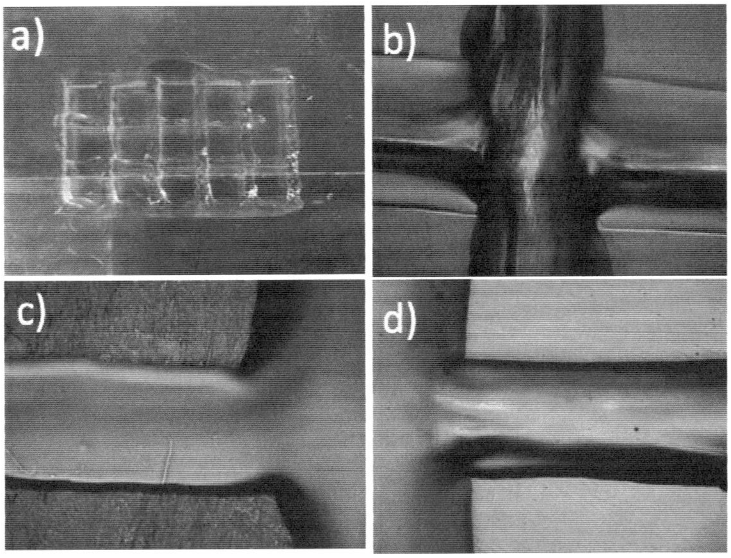

Figure 5. Camera image of gelatine structure and optical microscope images of the gelatine scaffolds taken at 5x magnification. a) gelatine scaffold b) 10 wt.% c) 15 wt.% d) 20 wt.%.

4. Conclusion

The gelatine solutions were printed succesfully via modified 3D printing device. It has been found that stifness of the scaffolds were highly dependent on viscosity, surface tension, temperature and concentration of the solutions. It can be seen that, 10 wt.% of gelatine sample was printed easily according to the lower viscosity. But 15 and 20 wt.% gelatine samples were accomplished much more smooth and uniform surface compared with the low concentrations.

Acknowledgement

This study is granted by Marmara University, BAPKO foundation (Project No: FEN-C-YLP-120417-0177) and Republic of Turkey Ministry of Development (Project No: 2016K121280).

References

[1] Schuurman W, Levett P, Pot M W , van Weeren P R, Dhert J A Wouter, Hutmacher W D, Melchels P W F, Klein J T and Malda J 2013 *Macromol. Biosci.* **13(5)** 551-61
[2] Ito A, Mase A, Takizawa, Shinkai M, Honda H, Hata K I, Ueda M and Kobayashi T 2003 *Biosci. Bioeng.* **95** 196-99
[2] Chia N H and Wu M B 2015 *J Biomed Eng* **9(1)** 4
[4] Luo Y, Lode A, Akkineni A R and Gelinsky M 2015 *RSC Advanc.* **5(54)** 43480-88
[5] Gupta A, Mohanty B and Bohidar H B 2005 *Biomacromolecules* **6(3)** 1623-27
[6] Tseng H J, Tsou T L, Wang H J and Hsu S H 2013 *J. Tissue Eng.* **7** 20-31

IOP Conf. Series: Materials Science and Engineering **293** (2017) 012002 doi:10.1088/1757-899X/293/1/012002

Effects of sintering temperature on electrical properties of sheep enamel hydroxyapatite

F Dumludag[1], O Gunduz[2,3], O Kılıc[3,4], B Kılıc[5] N Ekren[3,6], C Kalkandelen[3,7] and F N Oktar[3,8]

[1]Marmara University, Faculty of Arts and Sciences, Department of Physics, 34722 Kadıkoy, Istanbul / TURKEY
[2]Marmara University, Faculty of Technology, Department of Metallurgical and Materials Eng., 34722 Kadıkoy, Istanbul / TURKEY
[3]Marmara University, Advanced Nanomaterials Research Lab., 34722 Kadıkoy, Istanbul / TURKEY
[4]Marmara University, Faculty of Engineering, Department of Electrical and Electronics Eng., 34722 Kadıkoy, Istanbul / TURKEY
[5]Yıldız Technical University, Faculty of Electric-Electronic Engineering, Department of Electric Eng., 34220, Istanbul / TURKEY
[6]Marmara University, Faculty of Technology, Department of Electrical and Electronic Eng., 34722 Kadıkoy, Istanbul / TURKEY
[7]Istanbul University, Vocational School of Technical Sciences, Department of Biomedical Devices Technology, Istanbul / TURKEY
[8]Marmara University, Faculty of Engineering, Department of Bioengineering, 34722 Kadıkoy, Istanbul / TURKEY

fatihdumludag@marmara.edu.tr

Abstract. Bioceramics, especially calcium phosphate based bioceramics, whose examples are hydroxyapatite, and calcium phosphate powders have been widely used in the biomedical engineering applications. Hydroxyapatite (HA) is one of the most promising biomaterials, which are derived from natural sources, chemical method, animal like dental enamel and corals. The influence of sintering temperature on the electrical properties (i.e. DC conductivity, AC conductivity) of samples of sintered sheep enamel (SSSE) was studied in air and in vacuum ambient at room temperature. The sheep enamel were sintered at varying temperatures between 1000°C and 1300°C. DC conductivity results revealed that while dc conductivity of the SSSE decreases with increasing the sintering temperature in air ambient the values increased with increasing the sintering temperature in vacuum ambient. AC conductivity measurements were performed in the frequency range of 40 Hz - 10^5 Hz. The results showed that ac conductivity values decrease with increasing the sintering temperature.

IOP Conf. Series: Materials Science and Engineering **293** (2017) 012002 doi:10.1088/1757-899X/293/1/012002

1. Introduction

With increased living standards, an increase in injuries due to bone deterioration has also been observed. Therefore, it is very important to heal these injuries as soon as possible. To resolve this problem, surgical operation usually performed. But after that, ıt can take a very long time for bone to recover its mechanical properties in the first place [1]. Enamel is the most highly mineralized skeletal tissue present in the body. Mature enamel is consisted of 95-97 % carbonated Hydroxyapatite (HA) and 1% organic material [2]. HA is the most attractive as replacement of the inorganic material of bones. There has been attracted attention in the last years for the investigation of the electrical properties of HA material. Valerio and coworkers [3, 4], showed the positive effects of higher conductivity on bone healing for HA. Rodrigues et al. have examined piezoelectric properties because one of the mechanisms for bone growth is electrochemical process due to the generation of piezoelectric dipoles [5]. Electrical characterization is the powerful technique for investigation for some process such as dehydration and dehydroxylation in HA [6] and microstructure [7]. To date, a few study were reported on electrical characterization of SSSE. Because of the importance of electrical characterization of these materials there is a need to study of HA based material. In this study dc and ac electrical properties of the SSSE were investigated as a function of sintering temperature in air and in vacuum ambient, in dark and at room temperature.

2. Materials and experimental procedure

Extracted sheep teeth were obtained from local shop in Istanbul. The teeth were immersed in an antiseptic solution to avoid odour and possible contamination caught during preparation course. The teeth were cleaned slightly with continuous distilled water irrigation and then deproteinized in an alkali solution. After drying, calcined at 850 ^{0}C for nearly 4 h to remove any proteins and prions. Enamel was clearly peeled off from dentine. The obtained enamel samples were ground by a ball grinder for 4 h (Planetary Ball Mill PM 200 Retsch, Haan, Germany). Sheep enamel powders were sieved. The enamel powders were obtained nearly 100-150 μm. The prepared powders were pressed between hardened steel dies under the pressure of 350 MPa. The green sheep enamel compacts were sintered for 4 h at different temperatures (1000, 1100, 1200, and 1300 ^{0}C) (Nabertherm HT 16/17, Lilienthal, Germany). In order to prepare the pellets of SSSE for dc and ac electrical measurements circular-shaped silver electrodes on both surfaces of the SSSE were coated in high vacuum ambient (<10^{-6} mbar) using thermal evaporation method. DC and ac conductivity measurements were performed in aluminum chamber in air and in vacuum ambient (~$2x10^{-2}$ mbar) in dark at room temperature. During the dc conductivity measurements sweep voltage between -1 and 1 volt was applied to the SSSE and current passing through to SSSE was measured using Keithley model 6517B programmable electrometer. AC conductivity measurements were performed between the frequencies 40 Hz $1x10^5$ Hz using a Keithley model 3330 LCZ meter. All the measurement system was computerized.

3. Results and discussion

In order to determine effects of sintering temperature on dc and ac conductivity properties of the SSSE, electrical measurements were performed. Electrical conductivity measurements were done on the sheep enamel sintered at different temperatures between 1000 -1300 ^{o}C in air and in vacuum ambient (<10^{-2} mbar) in dark at room temperature. In order to calculate conductivity of the SSSE, current values were recorded between the applied voltages of -1 and 1 V. Current-Voltage (I-V) curves of the SSSE at indicated temperatures in air and in vacuum ambient were given in Figure 1. I-V curves of the SSSE showed ohmic behavior in air and in vacuum ambient. By inspecting the Fig. 1 hysteresis were observed in I-V curves except the SSSE at 1000°C and 1100°C in air ambient. The hysteresis may arises from deep traps as previously reported [7, 8]. This processes need longer time constant for charging and discharging process. Temperature, voltage sweeps speed and deep trap level may affects the hysteresis.

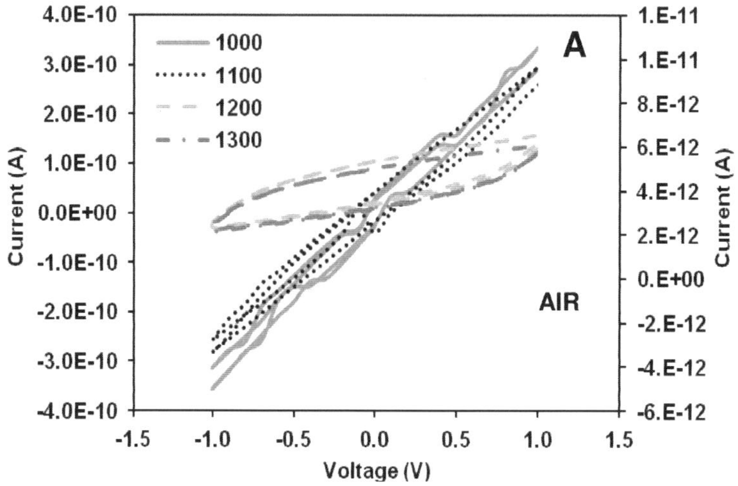

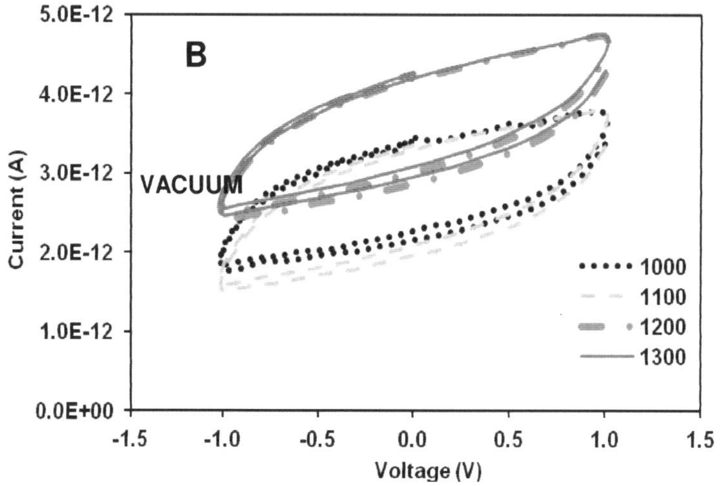

Figure 1. Current-Voltage (I-V) curves of the sheep enamel at indicated sintering temepratures obtained in dark (a) in air ambient (the SSSE at 1000 °C uses right-axis, and rest of the sheep enamel use left-axis) (b) in vacuum ambient.

DC conductivity values of the SSSE were calculated using the slopes of the I-V curves. The calculated d.c. conductivity values of the sheep enamel as a function of sintering temperatures in air and in vacuum ambient were presented in Figure 2. DC conductivity value was calculated as $9.77x10^{-11}$ S/m for the composite sintered at 1000 °C and increased to $3.01x10^{-10}$ S/m for the sheep enamel sintered at 1300 °C in vacuum ambient. By examining Fig. 2 effects of sintering temperature can be seen easily. First observation is that since the scale of vertical axis is logarithmic scale it can be concluded that sintering temperature affects strongly the dc conductivity. Another observation is the opposite behavior of dc conductivity in air and in vacuum ambient depending on sintering temperature. While dc conductivity decreases with increasing sintering temperature in air it increases with increasing sintering temperature in vacuum. The decrease in conductivity in air may arise from oxygen absorption. The real effect of the sintering temperature on dc conductivity of the sheep enamel can be obtained by eliminating another effects other than temperature such as humidity and oxygen. The conductivity of the sheep enamel increased with increasing sintering temperature in vacuum ambient. The behavior may be attributed to decrease in resistance in two ways: i) increasing sintering temperature decrease the particle size and that results the decrease in grain boundary resistance, ii) probably change in molecular structure with increasing sintering temperature that results the decrease in bulk resistance.

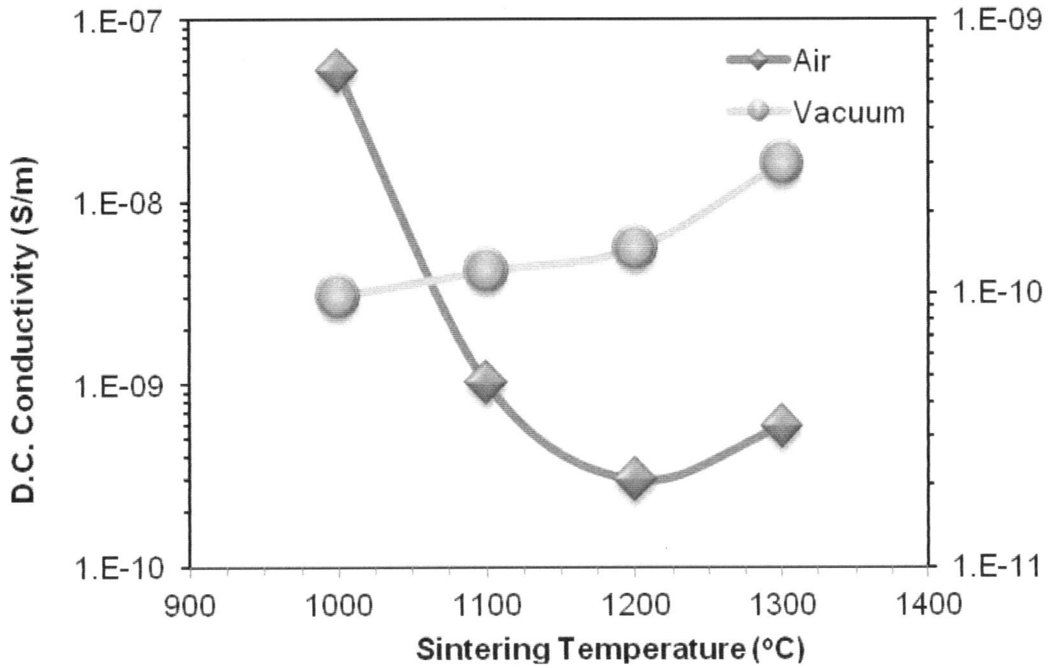

Figure 2. Variation of D.C. conductivity values as a function of sintering temperature in air and in vacuum.

AC conductivity measurements were also performed in air and vacuum ambient in the frequency range of 40 Hz - 10^5 Hz. AC conductivity values were decreased from $1.21x10^{-6}$ S/m to $3.69x10^{-7}$ S/m as sintering temperature increased from 1000 °C to 1300 °C at 10 kHz in air ambient. The values were decreased from $6.06x10^{-7}$ S/m to $2.46x10^{-7}$ S/m as sintering temperature increased from 1000 °C to 1300 °C at 10 kHz in vacuum ambient. It can be concluded that ac conductivity values decreases with increasing sintering temperature.

4. Conclusions

The present study of the influence of sintering temperature on the electrical properties of the calcinated sheep enamel, prepared from derived sheep teeth, observed that sintering temperature of the sheep enamel affected strongly the dc and ac conductivity of the SSSE. From conductivity measurements it can be concluded that, basically, dc conductivity of the sheep enamel directly proportional and ac conductivity of the sheep enamel inversely proportional to sintering temperature.

5. Acknowledgements

This study is granted by Marmara University, BAPKO foundation (Project No: FEN-C-YLP-120417-0177) and Republic of Turkey Ministry of Development (Project No: 2016K121280).

6. References

[1] F N Oktar 2006 *Mater. Lett.* **60** 2207-2210

[2] L C Palmer, C J Newcomb, S R Kaltz, E D Spoerke and S I Stupp 2008 *Chem. Tech.* **108(11)** 4754-4783

[3] A P M Schainberg, P Valerio, F N Oktar, M F Leite, A M Goes 2006 *20th European Conference on Biomaterials Proceedings* Nantes
[4] A P M Shainberg, P. Valerio, F N Oktar 2012 *Adv. Mater. Sci. Eng.* **2012** 1-10
[5] R Rodriguez, D Rangel, G Fonseca, M Gonzalez, S Vargas 2016 *Res. Phys.* **6** 925-32
[6] J P Gittings, C R Bowen, A C E Dent, I G Turner, F R Baxter, J B Chaudhuri *Acta. Biomater.* **5** 743-54
[7] C C Silva, M P F, Graça, M A Valente, A S B Sombra 2006 *J. Non. Cryst. Solids.* **352** 1490-94
[8] R Bayrak, F Dumludag, H T Akcay, I Degirmencioglu 2013 *Spectrochım Acta. A.* **105** 550-56
[9] P D Hooper, M I Newton, G McHale, M R Willis 1992 *Semicond. Sci. Technol.* **12** 455-59

IOP Conf. Series: Materials Science and Engineering **293** (2017) 012003 doi:10.1088/1757-899X/293/1/012003

DC and AC conductivity properties of bovine dentine hydroxyapatite (BDHA)

F Dumludag[1], O Gunduz[2,3], O Kılıc[3,4], N Ekren[3,5], C Kalkandelen[3,6], B Ozbek[3,7] and F N Oktar[3,8]

[1]Marmara University, Faculty of Arts and Sciences, Department of Physics, 34722 Kadıkoy, Istanbul / TURKEY
[2]Marmara University, Faculty of Technology, Department of Metallurgical and Materials Eng., 34722 Kadıkoy, Istanbul / TURKEY
[3]Marmara University, Advanced Nanomaterials Research Lab., 34722 Kadıkoy, Istanbul / TURKEY
[4]Marmara University, faculty of Engineering, Department of Electrical and Electronics Eng., 34722 Kadıkoy, Istanbul / TURKEY
[5]Marmara University, Faculty of Technology, Department of Electrical and Electronic Eng., 34722 Kadıkoy, Istanbul / TURKEY
[6]Istanbul University, Vocational School of Technical Sciences, Department of Biomedical Devices Technology, Istanbul / TURKEY
[7]Marmara University, Institute of Pure and Applied Sciences, Department of Metallurgical and Materials Eng., 34722 Kadıkoy, Istanbul / TURKEY
[7]Marmara University, Faculty of Engineering, Department of Bioengineering, 34722 Kadıkoy, Istanbul / TURKEY

foktar@marmara.edu.tr

Abstract. Bovine dentine bio-waste may be used as a potential natural source of hydroxyapatite (BDHA), thus extraction of bovine dentin hydroxyapatite (BDHA) from bio-waste is significantly important to fabricate in a simple, economically and environmentally preferable. DC and AC conductivity properties of BDHA were investigated depending on sintering temperature (1000°C - 1300°C) in air and vacuum ($<10^{-2}$ mbar) ambient at room temperature. DC conductivity measurements performed between -1 and 1 V. AC conductivity measurements performed in the frequency range of 40 Hz – 100 kHz. DC conductivity results showed that dc conductivity values of the BDHA decrease with increasing sintering temperature in air ambient. It is not observed remarkable/systematic behavior for ac conductivity depending on sintering temperature.

1. Introduction

Recently, hydroxyapatite (HA) has widely used as a crucial biomaterials due to brilliant properties of biocompatibility and high osteogenic potential. Its crystallographic and chemical properties closely resemble those of bone and tooth minerals [1]. Therefore HA is commonly used in in several hard tissue orthopedic applications. There has been attracted attention in the last years for the investigation of the electrical properties of HA. Valerio and coworkers [2, 3], presented their reports showing constructive contribution of higher conductivity on bone healing for HA composite. Another study were performed by Rodrigues and coworkers on piezoelectric properties of the composites since it is well known, that one of the mechanism for bone growth is electrochemical process due to the generation of piezoelectric dipoles [4]. Electrical characterization is very informative tool to study for some process such as dehydration and dehydroxylation in HA [5] and investigate variations in microstructure [6].

Up to now, a small number of reports were presented on electrical characterization of HA. Because of the importance of electrical characterization of these materials, since positive contribution of higher conductivity on bone growing, there is a need to study of HA based materials. In this study dc and ac electrical properties of the BDHA were investigated as a function of sintering temperature, in air and in vacuum ambient, in dark and at room temperature.

2. Materials and experimental procedure

Extracted bovine teeth were obtained from local shop in Istanbul. The teeth were immersed in an antiseptic solution to avoid bad smell and possible contamination caught during preparation course. The teeth were cleaned slightly with continuous distilled water irrigation and then deproteinized in an alkali solution. After drying process, teeth calcined at 850^{0}C for nearly 4 h to remove any proteins and prions. Dentine was clearly peeled off from enamel. The obtained dentine samples were ground by a ball grinder for 4 h (Planetary Ball Mill PM 200 Retsch, Haan, Germany). Sheep dentine powders were sieved. The dentine powders were obtained nearly 100-150 µm. The prepared powders were pressed between hardened steel dies under the pressure of 350 MPa. The green bovine dentine compacts were sintered for 4 h at different temperatures (1000, 1100, 1200, and 1300^{0}C) (Nabertherm HT 16/17, Lilienthal, Germany). Four pellets were prepared in order to investigate sintering temperature effects on BDHA. The prepared pellets of the composites were presented in Table 1. Silver electrodes were deposited onto both sides of pellets of BDHA in high vacuum ambient ($<10^{-6}$ mbar) using thermal evaporation method to prepare the composites to measure current passing through the pellets. DC and ac conductivity measurements were done on the composites in air and in vacuum ambient (~2x10^{-2} mbar) at room temperature. The measurements were performed in dark ambient to prevent the current from light effects. Current passing through the composites were measured using Keithley model 6517B programmable electrometer and recorded by applying voltages between the -1 V and 1 V with 50 mV steps to the terminals of the pellets. AC conductivity measurements were done on the composites between the frequencies 40 Hz - 100 kHz using Keithley model 3330 LCZ meter. All the measurement system was computerized.

Table 1. Prepared Bovine Dentine Hydroxyapatite (BDHA)

Label of the Bovine Dentine Hydroxyapatite	BDHA 1000	BDHA 1100	BDHA 1200	BDHA 1300
Sintering Temperature [^{o}C]	1000	1100	1200	1300

3. Results and discussion

Electrical measurements were performed in order to determine effects of sintering temperature on electrical properties of the BDHA. Electrical conductivity measurements were done in air and in vacuum ambient ($<10^{-2}$ mbar) in dark at room temperature on the BDHA sintered at the temperatures of 1000°C - 1100°C, 1200°C, and 1300°C. Conductivity of the BDHA were calculated using the slope of current - voltage (I-V) curves between the applied voltages of -1 and 1V. Figure 1 shows current - voltage (I-V) curves of the BDHA in air and in vacuum ambient sintered at indicated temperatures. I-V curves of the BDHA revealed ohmic behavior in air and in vacuum ambient. As can be seen from the Figure 1 hysteresis was observed in I-V curves obtained in air and in vacuum ambient. The hysteresis may arise from deep traps as previously reported [7, 8]. These processes need longer time constant for charging and discharging process. Temperature, voltage sweeps speed and deep trap level may affect the hysteresis.

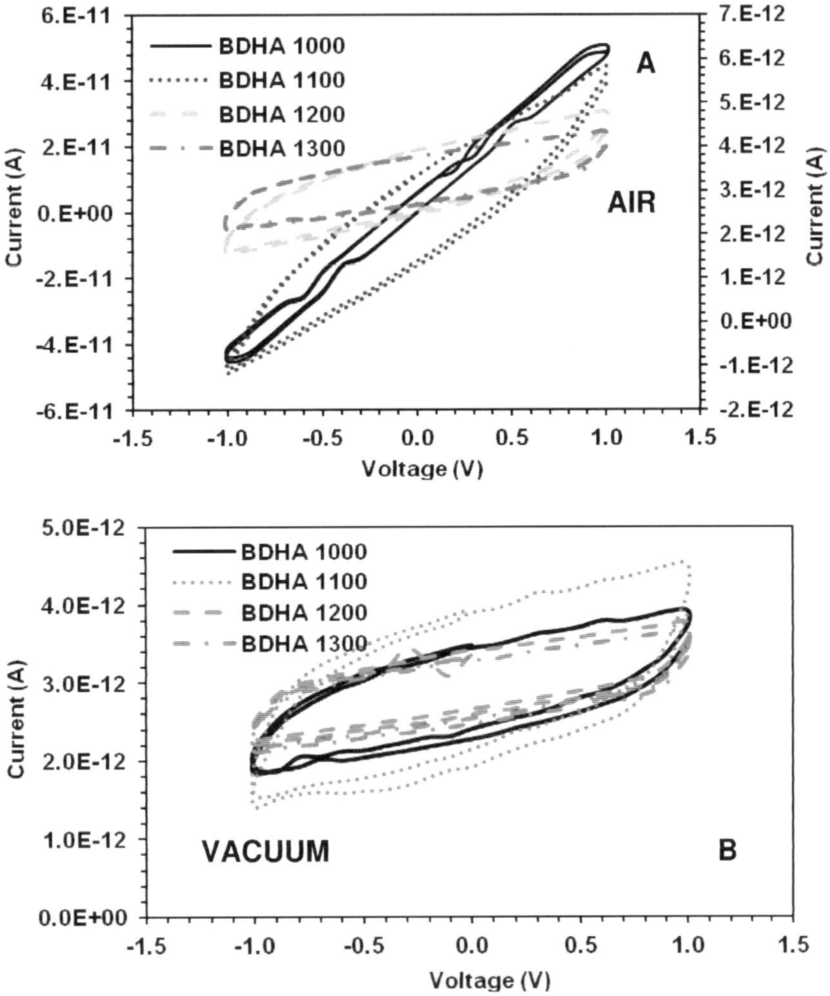

Figure 1. Current-Voltage (I-V) curves of the BDHA obtained in a dark (a) in air ambient (BDHA 1000 uses left-axis, and the rest of the BDHA use right-axis) (b) in vacuum ambient

Figure 2 shows the calculated d.c. conductivity values of the BDHA as a function of sintering temperatures in air and in vacuum ambient. DC conductivity value was calculated as 8.33×10^{-9} S/m for the BDHA1000 and decreased to 1.21×10^{-10} S/m for the BDHA 1300 in air ambient. In contrast to the behavior in air the values were fluctuated depending on sintering temperatures in the temperatures range of 1000°C - 1300°C. The values took place 6.53×10^{-11} S/m and 1.34×10^{-10} S/m. By inspecting Fig. 2 effects of sintering temperature can be seen easily. It might be said that since the scale of vertical axis is logarithmic scale dc conductivity values were strongly affected from sintering temperature. The effect of the sintering temperature on dc conductivity of the BDHA can be obtained more clearly by eliminating effects other than temperature such as humidity and oxygen. Because of this fact the measurements were also performed in vacuum ambient. The conductivity of the BDHA tends to decrease with increasing sintering temperature in vacuum ambient. The behavior may be explained as decrease in resistance in two ways: i) decreasing the particle size with increasing sintering temperature that results the decrease in grain boundary resistance, ii) increasing sintering temperature may small changes in molecular structure that results the decrease in bulk resistance.

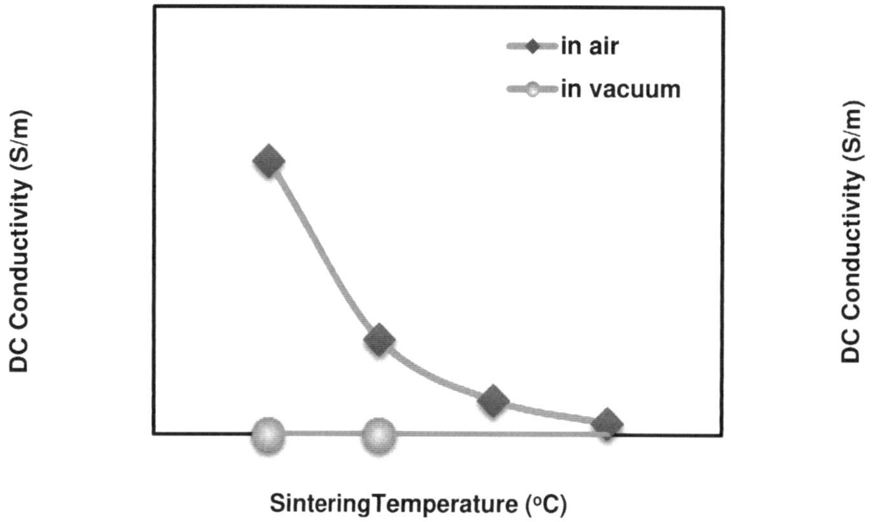

Figure 2. Variation of d.c conductivity values as a function of sintering temperature in air and in vacuum (dc conductivity in air uses left-axis, dc conductivity in vacuum use right-axis)

AC measurements were also done on the BDHA in air and vacuum ambient in the frequency range of 40 Hz – 100 kHz. AC conductivity values of the composites took placed between $5.90x10^{-7}$ S/m and $9.97x10^{-7}$ S/m at 10 kHz in air ambient. The values were decreased to the range of $5.88x10^{-7}$ S/m - $6.64x10^{-7}$ S/m at 10 kHz in vacuum ambient. Although there are some changes in ac conductivity depending on sintering temperatures but it is not observed remarkable/systematic behavior for ac conductivity depending on sintering temperature. This point needs to investigate in more detail such as ac conductivity variation as a function of frequency at elevated ambient temperatures in vacuum. The measurements are also underway.

4. Conclusions

This study shows the influence of varying sintering temperature on the DC and AC conductivity properties of the BDHA (prepared from extracted bovine dentine). It was investigated that sintering temperature of the bovine dentine were affected powerfully the dc conductivity of the samples of BHDA. From dc conductivity measurements it can be concluded that, although there is fluctuation in dc conductivity values of the samples in vacuum but in general it tends to decrease with increasing sintering temperature. It is not observed remarkable/systematic behavior for ac conductivity depending on sintering temperature.

Acknowledgements

This study is granted by Marmara University, BAPKO foundation (Project No: FEN-C-YLP-120417-0177) and Republic of Turkey Ministry of Devolopment (Project No: 2016K121280).

References

[1] H H Celik, O Gunduz, N Ekren, Z Ahmad and F N Oktar 2011 *J. Biomater. Nanobiotechnol.* **2** 98-101
[2] A P M Schainberg, P Valerio, F N Oktar, M F Leite, A M Goes 2006 *20th European Conference on Biomaterials Proceedings* Nantes
[3] A P M Shainberg, P. Valerio, F N Oktar 2012 *Adv. Mater. Sci. Eng.* **2012** 1-10
[4] R Rodriguez, D Rangel, G Fonseca, M Gonzalez, S Vargas 2016 *Res. Phys.* **6** 925-32
[5] J P Gittings, C R Bowen, A C E Dent, I G Turner, F R Baxter, J B Chaudhuri *Acta. Biomater.* **5** 743-54
[6] C C Silva, M P F, Graça, M A Valente, A S B Sombra 2006 *J. Non. Cryst. Solids.* **352** 1490-94
[7] R Bayrak, F Dumludag, H T Akcay, I Degirmencioglu 2013 *Spectrochim Acta. A.* **105** 550-56
[8] P D Hooper, M I Newton, G McHale, M R Willis 1992 *Semicond. Sci. Technol.* **12** 455-59

Synthesis and Property of Ag(NP)/catechin/Gelatin Nanofiber

Muhamad Nasir[1*] and Dita Apriani[2]

[1]Research Unit for Clean Technology, Indonesian Institute of Sciences (LIPI), Jalan Cisitu Sangkuriang, Bandung, West Java, Indonesia 40135
[2]Deparment of chemistry, Universitas Achmad Yani, Cimahi, West Java

*Corresponding Author: mnasir71@yahoo.com

Abstract. Nanomaterial play important role future industry such as for the medical, food, pharmaceutical and cosmetic industry. Ag (NP) and catechin exhibit antibacterial property. Ag(NP) with diameter around 15 nm was synthesis by microwaved method. We have successfully produce Ag(NP)/catechin/gelatin nanofiber composite by electrospinning process. Ag(NP)/catechin/gelatin nanofiber was synthesized by using gelatin from tuna fish, polyethylene oxide (PEO), acetic acid as solvent and silver nanoparticle(NP)/catechin as bioactive component, respectively. Morphology and structure of bioactive catechin-gelatin nanofiber were characterized by scanning electron microscopy (SEM) and fourier transform infrared spectroscopy (FTIR), respectively. SEM analysis showed that morphology of nanofiber composite was smooth and had average diameter 398.97 nm. FTIR analysis results were used to confirm structure of catechin-gelatin nanofiber. It was confirmed by FTIR that specific vibration band peak amide A (N-H) at 3286,209 cm^{-1}, amide B (N-H) 3069,396 cm^{-1}, amide I (C=O) at 1643,813 cm^{-1}, amide II (N-H and CN) at 1538,949 cm^{-1}, amide III (C-N) at 1276,789 cm^{-1}, C-O-C from polyethylene oxide at 1146,418 cm^{-1}, respectively. When examined to S. Aureus bacteria, Ag/catechin/gelatin nanofiber show inhabitation performance around 40.44%. Ag(NP)/catechin/gelatin nanofiber has potential application antibacterial medical application.

1. Introduction

Silver nanoparticle, Ag(NP) is important nanomaterial which shows antibacterial performance and used in in cosmetic, soap and textile application [1]. One versatile process to produce Ag(NP) is produced by using reducing agent with microwave assisted method [2]. Catechin is polyphenolic compound obtained from extract of gambier, leaf green tea, apple, red wine. Catechin is used for food, medical, pharmaceutical and cosmetic industry and showed antioxidant, anti-inflammatory, anti-tumor and anti-bacteria activities properties [3,4]. Gelatin is protein which derived from collagen. Gelatin has wide range application such as for food, biotechnology, medical, pharmaceutical and cosmetic industry [5].

Electrospinning is versatile method to produce nanofiber. Electrospinning process is used direct current (DC) high voltage electric to drive nanofiber formation from polymer solution as main driving force process [6]. Wide spectrum aspects of research and application of gelatin nanofiber are explored and studied by scientist in the world. Laha et al have studied drug release of crosslinked gelatin nanofiber mesh [7]. Lee et al prepared FGF-2 immobilized gelatin nanofibers for tissue engineering [8]. Huang et al studied proanthocyanidin-crosslinked gelatin nanofibers for drug delivering application [9]. Zhuang et al produce chitosan/gelatin nanofibers containing silver nanoparticles [10]. Jalaja et al have studied property and activity of gelatin graphene oxide on S. Aureus and E. colli bacteria [11]. Biodegradation of PCl/gelatin nanofiber was studied by Dulnik et al [12]. Sheath/core PVA/gelatin nanofiber was synthesized and studied by Merkle et al [13]. Xu et all develop fluoride-containing gelatin nanofiber scaffold [14]. Nasir et al has functionalization of gelatin nanofiber by catechin [15].

In current research, our focus is to synthesized Ag(NP)/bioactive/catechin/gelatin nanofiber by electrospinning and to study its morphology, nanostructure and antibacterial property.

2. Experiment

Materials and equipment. Silver nitrate (Merk), catechin from gambier (LIPI), gelatin from yellow finn skin tuna fish (LIPI), glacial acetic acid (Merck), polyethylene oxide, PEO, with molecular weight 600 kDa (Merck). All materials were used without further purification. Zuheros Nano computerized electrospinning machine (LIPI) was used to fabricate nanofiber.

Preparation of Ag(NP) solution

Sodium citrate (1 % w/v) was added and mixed to 10 mL of silver nitrate (1 mM) solution. Reaction of silver nanoparticle formation was conducted in 300 watts in microwave reactor for 5 minutes.

Electrospinning solution. Electrospinning solution was prepared as follow: a mixed of silver nanoparticle, Ag(NP) solution (58% v/v), catechin solution10% (v/v), gelatin 8% (w/v) and PEO 4 % (w/v) were dissolved in acetic acid 32% (v/v) and gently stirrer at room temperature for 24 hours.

Electrospinning process. Electrospinning process to synthesized nanofiber was carried out by using computerized electrospinning machine. Electrospinning solution which contain Ag(NP), catechin, gelatin and PEO was put into syringe and set up in electrospinning machine. Electrospinning parameters as follow: nozzle to collector distance (18 cm), applied voltage (14 kV), fixed polymer solution concentration, and flow rate (0.008 ml/m) were used to fabricate nanofiber

Morphology of Ag(NP)/catechin/gelatin nanofiber characterization. Morphology of electrospun gelatin nanofiber was characterized by SEM. After that, nanofiber diameter was calculated from SEM image of nanofiber by using image processing software (Adobe Photoshop software).

Structure of Ag(NP)/catechin/gelatin nanofiber characterization. Structure of Ag(NP)/catechin/gelatin nanofiber was characterized by using ATR-FTIR spectrophotometer (Nicolet). For measuring this method, thin film sheet of nanofiber was used.

Antibacterial performance of Ag(NP)/catechin/gelatin nanofiber. Antibacterial property of Ag(NP)/catechin/gelatin nanofiber was test to *staphylococcus aureus* bacteria by optical density method. The procedure as follow: 6 mL nutrient broth was added to test tube and the add 40 mL *staphylococcus aureus* and nanofiber sample. Incubation was conducted for 24 hours at 37 ^{o}C. Optical density of suspension was measured at 600 nm by using spectrophotometer (Agilent) (15).

3. Result and Discussion

Morphology of silver nanoparticle. Ag(NP) was synthesized by microwave method. Ag(NP) has light yellow color as shown in figure 1. Morphology and sized of nanoparticle was observed by transmittance electron microscopy (TEM). Figure 1b showed morphology of Ag(NP). Ag(NP) diameter is around 15 nm

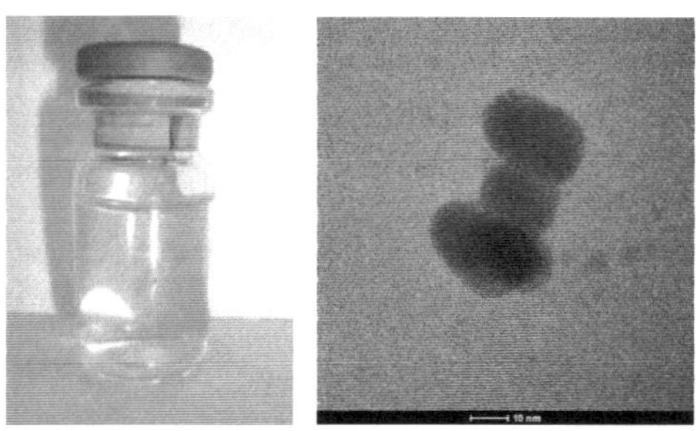

Figure 1. Morphology of silver nanoparticle, Ag(NP) analyzed by using TEM

Morphology of Ag(NP)/catechin/gelatin nanofiber. Ag/catechin/gelatin nanofiber was synthesized from mixed solution silver nanoparticle, catechin, gelatin and polyethylene oxide by using electrospinning process. PEO function as spinnability polymer for synthesis of nanofiber. After drying process, nanofiber was analyzed by SEM to obtain its morphology and diameter by using SEM. Morphology of nanofiber was smooth and

IOP Conf. Series: Materials Science and Engineering **293** (2017) 012004 doi:10.1088/1757-899X/293/1/012004

homogenous without bead nanofiber string. Range of nanofiber diameter was from 292 nm to 532 nm. The average diameter was 398.97 nm. Size of nanofiber diameter was influence by several factors such as polymer concentration, viscosity, surface tension applied voltage, distance between nozzle to collector and humidity [16].

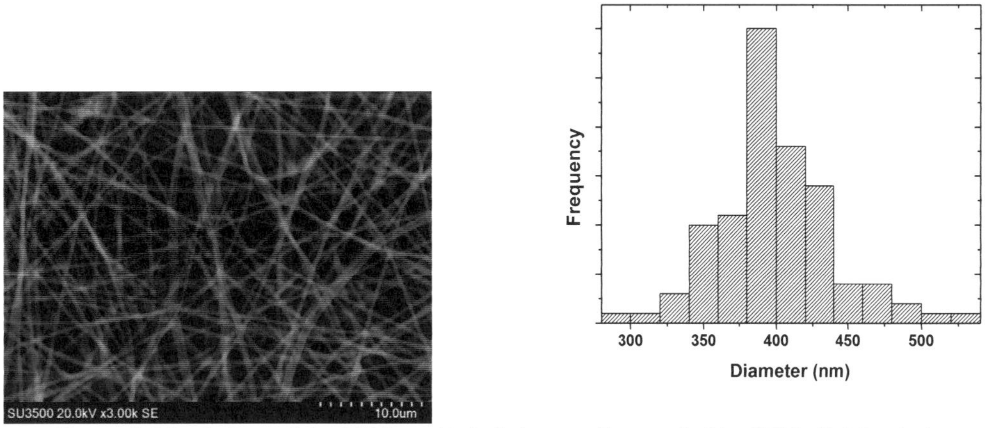

Figure 2. Morphology of Ag(NP)/catechin/gelatin nanofiber studied by SEM. A). Morphology of Ag/catechin/gelatin nanofiber, b). Distribution of fiber diameter

3.1. FTIR spectrum of Ag(NP)/catechin/gelatin nanofiber

Figure 3 showed FTIR spectra of Ag(NP)/catechin/gelatin nanofiber and catechin/gelatin nanofiber as control.

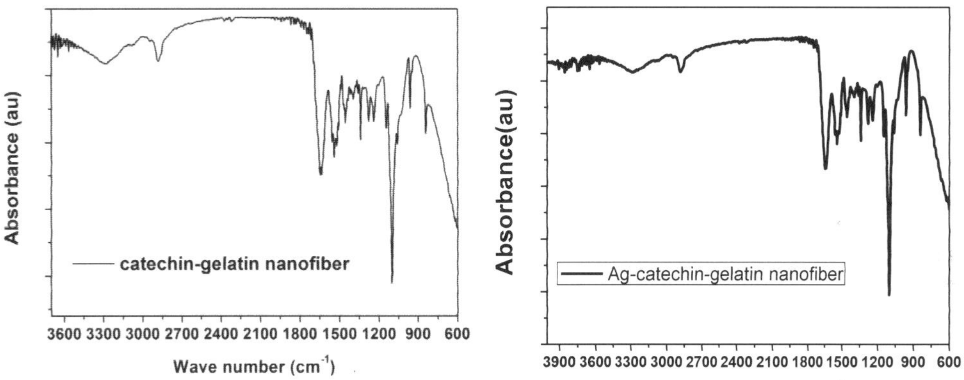

Figure 3. FTIR spectrum of nanofiber; a). Catechin/gelatin nanofiber [15] b). (a) and Ag(NP)/catechin/gelatin nanofiber

Figure 3 showed spectra of nanofiber. It was confirmed by FTIR that specific vibration band peak amide A (N-H) at 3286,209 cm^{-1}, amide B (N-H) 3069,396 cm^{-1}, amide I (C=O) at 1643,813 cm^{-1}, amide II (N-H and CN) at 1538,949 cm^{-1}, amide III (C-N) at 1276,789 cm^{-1}, C-O-C from polyethylene oxide at 1146,418 cm^{-1}, respectively. When we compare with our previous result in catechin/gelatin nanofiber [15], presence of Ag(NP) in nanofiber only slightly change of vibration peak of functional group gelatin, catechin and PEO but not change its structure.

3.2. Antibacterial Performance of Ag(NP)/catechin/gelatin nanofiber on bacteria

Ag(NP)/catechin/gelatin nanofiber antibacterial activity was examined to staphylococcus aureus bacteria by optical density/turbiditimetry methods. Staphylococcus aureus is well known as dangerous a gram positive of bacteria. S. Aureus bacteria is found on skin, hair, throat and nose.

A B

Figure 4. Appearance of antibacterial suspension samples. a). Control and b). Ag(NP)/catechin/gelatin nanofiber.

Appearance of suspension of test solution results were shown in figure 4. Bioactive catechin-gelatin nanofiber can inhibit S. Aureus bacteria around 40.44%. Gopal et al and Staphleton et al showed green tea of catechin effective in inhibiting bacteria. According to study of Grigor'eva et all, interaction between Ag(NP) with S. Aureus as follow: Ag(NP) adsorbed on cell of bacteria, penetrated and accumulated in cell walls and damage cytoplasm. Cytoplasm of bacteria become amorphous. Ag(NP) was bonded in DNA fiber S. Aureus [16,17,18 and 19].

4. Summary

We have successfully synthesized Ag/catechin-gelatin nanofiber with diameter around 398.97 nm by electrospinning process. Ag(NP)/catechin/gelatin nanofiber can inhibit growth of S. Aureus bacteria around 40.4%. Advanced study interaction of Ag(NP)/catechin with S. Aureus is required to explain the performance of its antibacterial activity. Ag(NP)/catechin-gelatin nanofiber has potential application for medical application.

Acknowledgements
This research is financial support by Ministry of Research and Technology-Higher Education Republic Indonesia (Insinas Project FY 2018-2019)

References

[1] B. Khodashenas and H. R. Ghorbani, "Synthesis of silver nanoparticles with different shapes," Arab. J. Chem., Jan. 2015.
[2] X. Zhao, Y. Xia, Q. Li, X. Ma, F. Quan, C. Geng, and Z. Han, "Microwave-assisted synthesis of silver nanoparticles using sodium alginate and their antibacterial activity," Colloids Surfaces A Physicochem. Eng. Asp., vol. 444, pp. 180–188, Mar. 2014.
[3] P. V. Gadkari and M. Balaraman, "Catechins: Sources, extraction and encapsulation: A review," *Food Bioprod. Process.*, vol. 93, pp. 122–138, Jan. 2015.
[4] R. K. Harwansh, P. K. Mukherjee, A. Kar, S. Bahadur, N. A. Al-Dhabi, and V. Duraipandiyan, "Enhancement of photoprotection potential of catechin loaded nanoemulsion gel against UVA induced oxidative stress," *J. Photochem. Photobiol. B Biol.*, vol. 160, pp. 318–329, Jul. 2016.
[5] Z. A. N. Hanani, "Gelatin," in *Encyclopedia of Food and Health*, 2016, pp. 191–195.
[6] M. Nasir, H. Matsumoto, T. Danno, M. Minagawa, T. Irisawa, M. Shioya, and A. Tanioka, "Control of diameter, morphology, and structure of PVDF nanofiber fabricated by electrospray deposition," *J. Polym. Sci. Part B Polym. Phys.*, vol. 44, no. 5, pp. 779–786, Mar. 2006.
[7] A. Laha, C. S. Sharma, and S. Majumdar, "Sustained drug release from multi-layered sequentially crosslinked electrospun gelatin nanofiber mesh," *Mater. Sci. Eng. C*, vol. 76, pp. 782–786, 2017.
[8] H. Lee, S. Lim, M. S. Birajdar, S.-H. Lee, and H. Park, "Fabrication of FGF-2 immobilized electrospun gelatin nanofibers for tissue engineering," *Int. J. Biol. Macromol.*, vol. 93, pp. 1559–1566, Dec. 2016.
[9] C.-H. Huang, C.-Y. Chi, Y.-S. Chen, K.-Y. Chen, P.-L. Chen, and C.-H. Yao, "Evaluation of proanthocyanidin-crosslinked electrospun gelatin nanofibers for drug delivering system," *Mater. Sci. Eng. C*, vol. 32, no. 8, pp. 2476–2483, Dec. 2012.
[10] X. Zhuang, B. Cheng, W. Kang, and X. Xu, "Electrospun chitosan/gelatin nanofibers containing silver nanoparticles," *Carbohydr. Polym.*, vol. 82, no. 2, pp. 524–527, Sep. 2010.

[11] K. Jalaja, V. S. Sreehari, P. R. A. Kumar, and R. J. Nirmala, "Graphene oxide decorated electrospun gelatin nanofibers: Fabrication, properties and applications," *Mater. Sci. Eng. C*, vol. 64, pp. 11–19, Jul. 2016.
[12] J. Dulnik, P. Denis, P. Sajkiewicz, D. Kołbuk, and E. Choińska, "Biodegradation of bicomponent PCL/gelatin and PCL/collagen nanofibers electrospun from alternative solvent system," *Polym. Degrad. Stab.*, vol. 130, pp. 10–21, Aug. 2016.
[13] V. M. Merkle, P. L. Tran, M. Hutchinson, K. R. Ammann, K. DeCook, X. Wu, and M. J. Slepian, "Core–shell PVA/gelatin electrospun nanofibers promote human umbilical vein endothelial cell and smooth muscle cell proliferation and migration," *Acta Biomater.*, vol. 27, pp. 77–87, Nov. 2015.
[14] J. Xu, J. Yan, Q. Gu, J. Li, and H. Wang, "Preparation of fluoride-containing gelatin nanofiber scaffold," *Mater. Lett.*, vol. 65, no. 15–16, pp. 2404–2406, Aug. 2011.
[15] M. Nasir and D. Apriani, "Synthesis of Catechin-Gelatin Nanofiber by Electrospinning," *Mater. Sci. Forum*, vol. 887, pp. 96–99, Mar. 2017.
[16] [R. Khajavi and M. Abbasipour, "5 – Controlling nanofiber morphology by the electrospinning process," in *Electrospun Nanofibers*, 2017, pp. 109–123.
[17] J. Gopal, M. Muthu, D. Paul, D.-H. Kim, and S. Chun, "Bactericidal activity of green tea extracts: the importance of catechin containing nano particles," *Sci. Rep.*, vol. 6, no. 1, p. 19710, Apr. 2016.
[18] P. D. Stapleton, S. Shah, K. Ehlert, Y. Hara, and P. W. Taylor, "The beta-lactam-resistance modifier (-)-epicatechin gallate alters the architecture of the cell wall of Staphylococcus aureus," *Microbiology*, vol. 153, no. 7, pp. 2093–2103, Jul. 2007.
[19] A. Grigor'eva, I. Saranina, N. Tikunova, A. Safonov, N. Timoshenko, A. Rebrov, and E. Ryabchikova, "Fine mechanisms of the interaction of silver nanoparticles with the cells of Salmonella typhimurium and Staphylococcus aureus," *BioMetals*, vol. 26, no. 3, pp. 479–488, Jun. 2013.

IOP Conf. Series: Materials Science and Engineering **293** (2017) 012005 doi:10.1088/1757-899X/293/1/012005

Pragmatic setup for bioparticle responses by dielectrophoresis for resource limited environment application

Mohd Anuar Md Ali[1], Burhanuddin Yeop Majlis[1], Aminuddin Ahmad Kayani[2*]

[1]Institute of Microengineering and Nanoelectronics, Universiti Kebangsaan Malaysia, Bangi, Selangor 43600, Malaysia
[2]Faculty of Engineering and Technology, Multimedia University, Melaka, Melaka 75450, Malaysia

*aminuddin.kayani@mmu.edu.my

Abstract. Various dielectrophoretic responses of bioparticles, including cell-chain, spinning, rotation and clustering, are of high interest in the field due to their benefit into application for biomedical and clinical implementation potential. Numerous attempts using sophisticated equipment setup have been studied to perform those dielectrophoretic responses, however, for development into resource limited environment application, such as portable, sustainable and environmental friendly diagnostic tools, establishment of pragmatic setup using standard, non-sophisticated and low-cost equipment is of important task. Here we show the advantages in the judicious design optimization of tip microelectrode, also with selection of suspending medium and optimization of electric signal configuration in establishing setup that can promote the aforementioned dielectrophoretic responses within standard equipments, i.e. pragmatic setup.

1. Introduction

Development of diagnosis tools for resource-limited environment such as war-torn and natural disaster situation are among major studies implementing dielectrophoresis research. While the development is in extensive progress with various sophisticated instrumentation and setup, the need for a pragmatic setup, i.e. portable, sustainable, environmental friendly is often overlooked.

Dielectrophoretic response of particles relies on their dielectric constant as well as of the suspending medium, and electric signal [1]. While the particles properties are of the major subject of the studies in this field and non-changeable parameter, the suspending medium and electric signal configurations are customizable.

Researchers can choose parameter for suspending medium, by selecting such as culture medium [2], sugar alcohol [3] and salty solution such as phosphate buffered saline (PBS) [4], or deionized water (DIW). However, for a portable, sustainable and environmental friendly selection for a point-of-care (POC) or war-torn resource limited and quick diagnostic tool, DIW is of the best choice. As the process of changing the suspending medium typically is centrifuge based process, which is simple, low cost and portable, the use of DIW as suspending medium is not labour-intensive, i.e practical option. In fact, DIW suspended bioparticles remain viable, at least for a period up to an hour as demonstrated in this work, which is sufficient enough for a quick diagnostic test.

Selection of electic signal parameters, i.e. the voltage, and frequency should be based on a standard function generator configuration so that the setup is non-sophisticated, portable and low cost. A standard function generator can produce a signal of voltage up to around tens V_{pp} and frequency range up to several MHz. There are a number of attempts to use sophisticated configuration such as GHz frequencies range, however this configuration might cost a larger space, less portability and higher expenses [5,6]. We showed that frequency range up until several MHz, which producible by a standard function generator, is sufficient enough to induce various responses of bioparticle, i.e. repel, attract, spin, rotate, and chaining. One should note that at frequencies less than 1 kHz, electro-osmosis and electrolysis could disrupt the cellular motional responses. Therefore, the ideal frequency ranges for bioparticles to exhibit dielectrophoretic responses in DIW is between tens of kHz to several MHz, as demonstrated in this work.

However, on top of mentioned parameters, the design of of the microelectrode design is the topmost important parameter. The best design is the one which can construct the maximum difference of the electric field gradient difference within the suspension region. In this work, we demonstrated a judicious design of microelectrode where microtip configuration is being employed, as it constructs non-continuous high electric field spots which can focus the bioparticle at positive dielectrophoretic response region (described in next section), instead of continuous high electric field region which reduce the contrast response of bioparticle between opposite responses (positive DEP and negative DEP).

We performed this work using widely used model organism, *Saccharomyces cerevisiae*, a type of yeast cell, which is a well-known and widely used in proof-of-concept studies, prior to expansion to study of various cells of interest.

Well-customized parameters in this work serves as a pragmatic setup to achieve bioparticle responses especially for the development of quick diagnostic tools for resource-limited environment, while at the same time open for future work using other targeted cells.

2. Dielectrophoresis

Dielectrophoresis is a motional phenomenon of polarizable dielectric materials, including bioparticles when suspended in a non-uniform electric field (Figure 1). Bioparticles are polarized, establishing dipoles within their membrane. Unequal net Coulombic forces are generated at opposite sides, pushing the bioparticles either towards or away from regions of electric field with high gradients. The acting force is known as dielectrophoretic (DEP) force, F_{DEP}, which is governed by [7,8]

$$F_{DEP} = 2\pi a^3 \varepsilon_m Re[f_{CM}]\nabla|E|^2 \quad (1)$$

where a is the radius of suspended particle or bioparticle, ε_m is the absolute permittivity of the suspending medium, f_{CM} is the Clausius–Mossotti factor and E is the electric field. The f_{CM} indicates the correlation between dielectric constants of two different media. For a spherical bioparticle, f_{CM} is given by

$$f_{CM} = \frac{\varepsilon_p^* - \varepsilon_m^*}{\varepsilon_p^* + 2\varepsilon_m^*} \quad (2)$$

where ε_p* and ε_m* are the complex permittivities of the suspended bioparticle and the suspending medium, respectively. The complex permittivity is determined by

$$\varepsilon^* = \varepsilon_0 \varepsilon - j\frac{\sigma}{\omega} \quad (3)$$

where, ε_0 is vacuum permittivity (8.854×10^{-12} F/m), ε is permittivity while σ is conductivity, j is $\sqrt{-1}$, and ω is angular frequency of supply voltage. The polarity of the DEP force depends on the sign of the real part of CM factor, Re[f_{CM}]. Positive Re[f_{CM}] indicates positive DEP (pDEP) force, pushing the bioparticles towards regions of high electric field gradients, while negative Re[f_{CM}] indicates negative DEP (nDEP) force, pushing bioparticles in opposite direction. The pDEP and nDEP responses of bioparticles alternate vice versa at a frequency known as crossover frequency, f_{xo}, where the polarizabilities of the bioparticles and the suspending medium are equal.

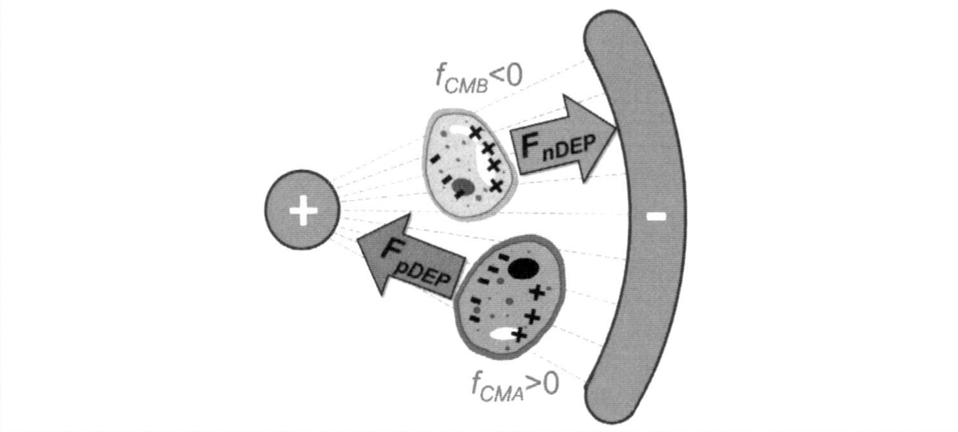

Figure 1. Dielectrophoretic response. Bioparticle with more polarizability against suspending medium is pushed (pDEP) to region with higher electric field gradient, while less polarizable bioparticle is pulled (nDEP) away [9].

3. Experimental

3.1. Computations and design.
Real part of the Clausius-Mossoti factor, Re[f_{CM}] is computed using Octave 4.0 for a frequency range between 1 kHz to 10 MHz. Electric field distributions E is simulated using Agros2D.

3.2. Device fabrication.
DEP platform consists of a glass microscope slide substrate with patterned DEP microtip electrodes, fabricated using standard photolithography and wet etching. The electrodes is structured by a 50 nm chromium (Cr) underlayer and 150 nm gold (Au) overlayer. A thick PDMS (DowCorning Corp, Sylgard 184) superstrate, formed using standard soft-lithography, consists of an inlet port, microchannel (150 μmW × 50 μmH) and outlet stage, and is bonded by corona plasma bonding to the top of the glass substrate.

3.3. Sample preparation.
Sample composed of yeast suspended in DIW with a volumetric ratio of 0.12% w/v, is capillary force driven into into the microchannel. The yeast suspension was sonicated in an ultrasonic bath at 37 °C for 30 min. The sample suspension is introduced into the microfluidic channel followed by a buffer period of 30 min prior to DEP response observation.

3.4. DEP response observation.
The DEP platform is mounted on the stage of an optical microscope (Motic) and the image is monitored by a monitor. Electric signal is applied to the electrodes by a function generator (Topward, 8120) while the signal configuration, i.e. voltage and frequency are monitored with an oscilloscope (Tektronix, TDS1012B). The voltage and frequency amplitudes are set at 7 V_{rms} (20 V_{pp}) and the frequency is varied between 20 kHz and 2 MHz.

4. Results and discussions

4.1. Simulation result
Simulation result of Re[f_{CM}] indicates that yeast cells suspended in DIW experience nDEP force at low AC frequencies, while pDEP force is experienced at higher frequencies, and cross over frequency takes place at f_{xo}=90 kHz (Figure 2).

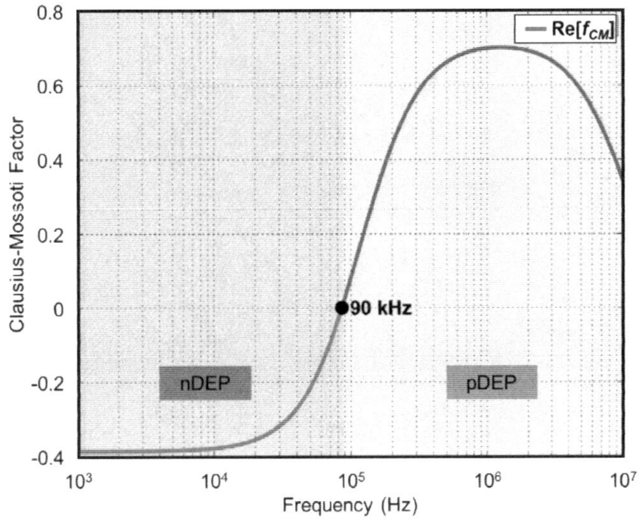

Figure 2. Simulation of Re[f_{CM}] for yeast cells (multi-shell model) suspended in DIW. Crossover frequency, f_{xo} = 90 kHz. Negative DEP (nDEP) is experienced at $f < f_{xo}$, while positive DEP (pDEP) occurs at $f > f_{xo}$.

Simulated electric field distribution (Figure 3) shows that highest electric field regions (230 kV/m) exist at the edge of microelectrode tips, i.e. 6 points (x = ±100 and 0 μm of y = ±20 μm), while lowest electric field regions exist at vicinity of the microelectrode base, i.e. along y = ±100 μm that without microtip.

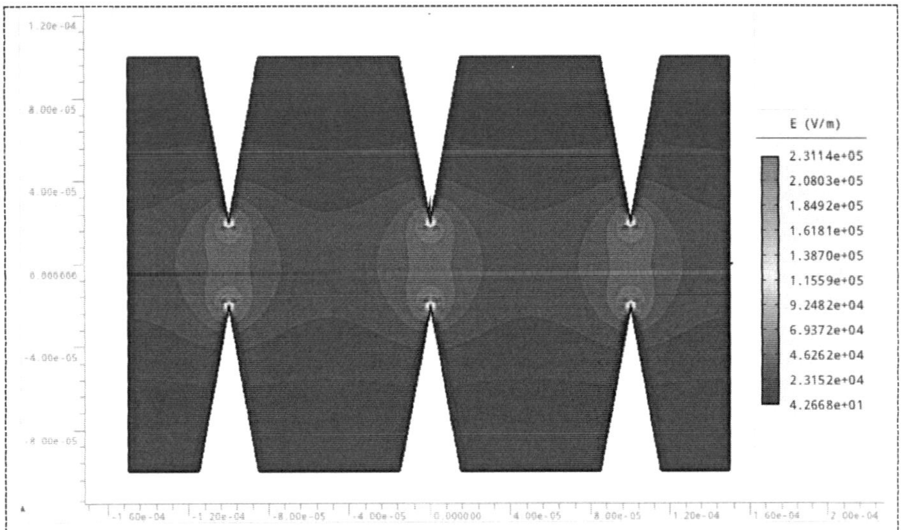

Figure 3. Simulation of electric field distribution, *E*.

4.2. Bioparticle responses

Observation of the yeast solution under bright field microscopy prior to the experiment confirms that the cells were in equilibrium and well dispersed before the application of the DEP force (Figure 4a).

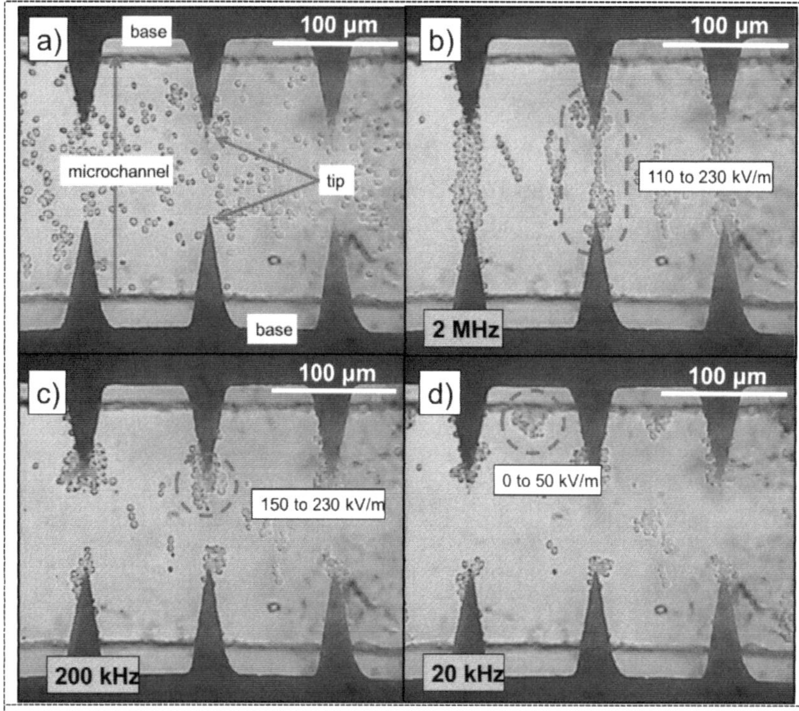

Figure 4. Bioparticle contact configurations under DEP force; a) prior application of AC signal, b) pDEP response at 2 MHz, c) pDEP response at 200 kHz, d) nDEP response at 20 kHz.

Under this pragmatic setup, cells can demonstrate the forming of chain structures with high frequencies (2 MHz) application. The cells at first move towards the microtip electrode at which the field intensity E is the maximum due to pDEP forces, and subsequently align into chain-like formation. The chains are formed throughout the microchannel while moving to finally localized at the centerline between the microelectrode pairs with cells fill in the gap to complete the bridging (Figure 4b). The chains spans a wide range of electric field intensities, from 230 kV/m as maximum located at the microtips edge to 110 kV/m as minimum located at the midpoint between the opposing microtip electrodes pairs. Establishment of the cells chain forms a stable contact between cells, promising for future application into studies of intercellular communication, bioelectricity and electrofusion.

Spinning of cells and rotation of cell clusters can be generated in addition to cells attract-clustering at 200 kHz frequency, not yet crossing the crossover f_{xo} (90 kHz). This frequency attracts the cells to accumulate at the microtip vicinity by pDEP force while at the same time the cells exhibiting spinning and rotational motion. The cells congregate close around the microtip edge with the electric field intensity ranges between 230 to 150 kV/m (Figure 4c). The characteristic of cells spinning and clusters rotational motion is cells dependent, hence promising for application into development for diagnostic tools and cells properties database.

In contrast to demonstration of cell-chaining, spinning, rotation and attract-clustering, the assembly of cells by repel-clustering can be formed at frequencies lower than the crossover frequency f_{xo}, which demonstrated at 20 kHz in this work. The cells are pulled away from regions of high field intensity in the vicinity of the microtip electrodes due to nDEP force to regions of low field intensity between 10 and 50 kV/m close to the microchannel side wall or the base regions (Figure 4d). Unlike the attract-clustering demonstrated in the pDEP response, cells in these nDEP repel-clustering exhibit a stable contact between the cells, i.e. do not spin and rotate, which expected to have potential application into studies based on as cell-chain formation such as cellular communication, however, with advantage in multi-directional configuration instead of linear configuration by cell-chain format..

5. Conclusions

Various bioparticles responses to DEP mechanisms are of high interest into application to development of biomedical and clinical studies. The different type of responses produced under pragmatic setup consist of customization of microtip electrode design, suspending medium selection and different standard electrical frequencies for resource limited environment have been discussed. Lower frequencies below the crossover frequency provides cell repel-clustering in low electric field intensity regions. Whereas attract-clustering can be obtained at higher frequency just above cross over frequency, with additional specific response in the form of cell spinning and cluster rotation. The construction of cell-chain structures can be generated at higher frequencies, i.e around MHz frequencies.

This pragmatic setup to induce various bioparticles responses is of highly potential implementation into studies in cellular communication and bioelectricity. As cell contact can be established at certain frequencies, how cell to cell communicate can be done by forming cell contact and later perform the desired analysis. As for cell spinning and cluster rotation, the characteristics such as relation of spinning speed and frequency is cell type dependent. Thus, characterization of various cell type especially disease related cells such as cancerous cell can be performed and later database could be established for disease diagnosis protocol in future. The development of a quick diagnostic tools for resource limited environment should get most benefit from this pragmatic setup optimization.

Acknowledgements

We thank the Ministry of Higher Education of Malaysia (MOHE) for support under Fundamental Research Grant Scheme FRGS/1/2015/TK04/MMU/02/9 and the Ministry of Education of Malaysia (MOE) for support under the Higher Institution Centre of Excellence (HiCOE) Grant, AKU-95.

References

[1] Çetin B, Özer M B and Solmaz M E 2014 Microfluidic bio-particle manipulation for biotechnology *Biochem. Eng. J.* **92** 63–82

[2] Mohammad K, Buchanan D A, Braasch K, Butler M and Thomson D J 2017 CMOS single cell dielectrophoresis cytometer *Sensors Actuators, B Chem.* **249** 246–55

[3] Agarwal S, Sebastian A, Forrester L M and Markx G H 2012 Formation of embryoid bodies using dielectrophoresis *Biomicrofluidics* **6** 024101–024101-11

[4] Cheng I F, Huang W L, Chen T Y, Liu C W, Lin Y D and Su W C 2015 Antibody-Free Isolation of Rare Cancer Cells from Blood based on 3D Lateral Dielectrophoresis *Lab Chip* **15** 2950-9

[5] Nikolic-Jaric M, Romanuik S F, Ferrier G A, Cabel T, Salimi E, Levin D B, Bridges G E and Thomson D J 2012 Electronic detection of dielectrophoretic forces exerted on particles flowing over interdigitated electrodes *Biomicrofluidics* **6** 024117–024117-15

[6] Stanke S, Bier F F and Hölzel R 2011 Fluid streaming above interdigitated electrodes in dielectrophoresis experiments *Electrophoresis* **32** 2448–55

[7] Kayani A A, Chrimes A F, Khoshmanesh K, Sivan V, Zeller E, Kalantar-Zadeh K and Mitchell A 2011 Interaction of guided light in rib polymer waveguides with dielectrophoretically controlled nanoparticles *Microfluid. Nanofluidics* **11** 93–104

[8] Çetin B and Li D 2011 Dielectrophoresis in microfluidics technology *Electrophoresis* **32** 2410–27

[9] Md Ali M A, Ostrikov K (Ken), Khalid F A, Majlis B Y and Kayani A A 2016 Active Bioparticle Manipulation in Microfluidic Systems *RSC Adv.* **6** 113066–94

IOP Conf. Series: Materials Science and Engineering **293** (2017) 012006 doi:10.1088/1757-899X/293/1/012006

Development of the Technology of Large Bodies Manufacturing Based on Combined Process of Plate Rolling and Stamping

Alexander Pesin[1], Ernst Drigun[2], Denis Pustovoitov[1], Ilya Pesin[1]

[1]*Nosov Magnitogorsk State Technical University, 38, Lenin pr., Magnitogorsk, 455000, Russia*
[2]LLC "ChermetInformSystemy", *11-54, Metallurgov pr., Magnitogorsk, 455000, Russia*

Corresponding author – Alexander Pesin pesin@bk.ru

Abstract. An experimental study of the developed new technologies for obtaining large-sized parts with an elliptical surface has been carried out. The advantages and disadvantages are established, as well as the boundaries of effective applicability of new combined process. Verification of the results of computer simulation based on testing the technology in the conditions of a laboratory rolling mill was performed. Optimal geometric shapes of matrices and punches to produce spherical and elliptical bottoms of various thicknesses and diameters are obtained on a thick plate rolling mill. The methods of correction of the technology for obtaining large-sized parts with an elliptical surface have been developed and tested in laboratory conditions to eliminate the identified shortcomings.

1. Introduction

To produce large-sized (4 ... 100 mm) complex profile details with specified curvature, a new combined thick-plate rolling and stamping process has been developed. It is based on the principle of a package rolling consisting of the upper (punch), the lower (matrix) bases and a thick plate between them. The production of long-dimensioned large bodies of specified curvature on a thick plate rolling mill opens the prospect of obtaining, for example, such products as casings of various large objects. It makes the work relevant for machine building it allows to obtain blanks for long-dimensioned large-sized bodies of specified curvature that cannot be obtained without welding with the help of press equipment [1-3]. To achieve this goal, the following tasks were set and solved: the development of a methodology for the experimental study of combined rolling and stamping processes; mathematical modelling and analysis of metal shaping and force parameters of a new process under conditions of implementation on a laboratory mill; a comparative analysis of merits and identification of shortcomings of direct and reversible schemes for implementing a combined process for obtaining various details such as bottoms; the development of corrective technical and technological impacts to eliminate previously identified shortcomings. The work was based on mathematical modelling and experimental research of a new process [4-6].

2. Experimental study of the combined process of rolling and stamping

To implement the new combined rolling and stamping process, a package consisting of the upper punch and the lower die was produced. The punch was provided with a ring-shaped projection extending radially inwards and having a convex surface. The lower matrix was made in the form of a ring 1.5 mm thick with an upper internal diameter of 160 mm. Between the punch and the matrix a sheet blank was placed made of lead 6 mm thick, which had a circle shape in the horizontal plane with a diameter of 183 mm (Fig. 1). The use of lead as a workpiece material makes it possible to simulate hot deformation at room temperature.

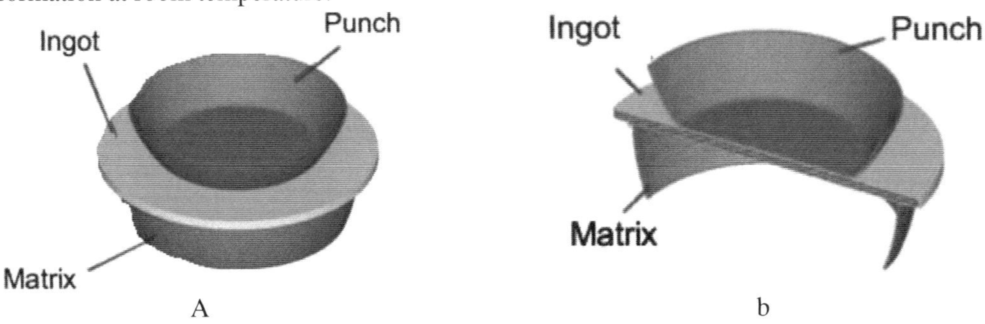

Figure 1. Experimental package: a) general view; b) cross section.

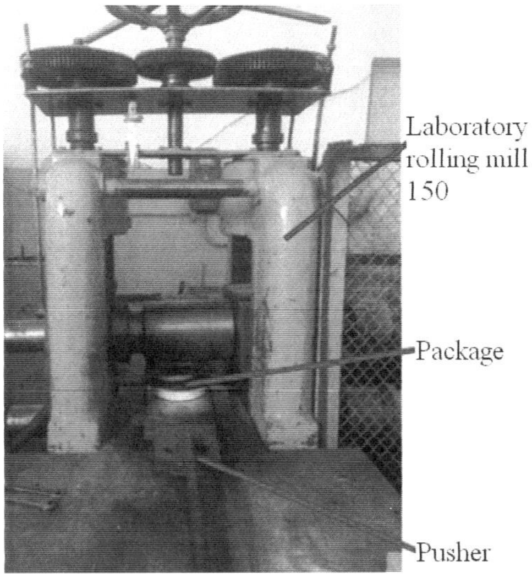

Figure 2. Laboratory rolling mill duo 150 for the implementation of the combined process "rolling-stamping"

Figure 3. Experimental package consisting of the upper (punch) and the lower (matrix) bases and the sheet blank between them.

The combined rolling and punching process was performed in one or several passes (from 2 to 4) in a straightforward manner. The accuracy of the resulting geometry of the product was evaluated based on the results of the experiment. In total, 10 items have been obtained. The curvature of the finished products was checked by special templates. The deviation was from 10 to 2%, depending on the number of passes. The more the number of passes, the less the deviation. The form of the final product of the given shape is shown in Fig. 4.

In the experiment, the influence of various friction on the quality of the resulting geometry of the finished product has also been investigated. It is shown that the greater the coefficient of friction at the punch-workpiece contact and the matrix-workpiece, the smaller the deviation of the geometry.

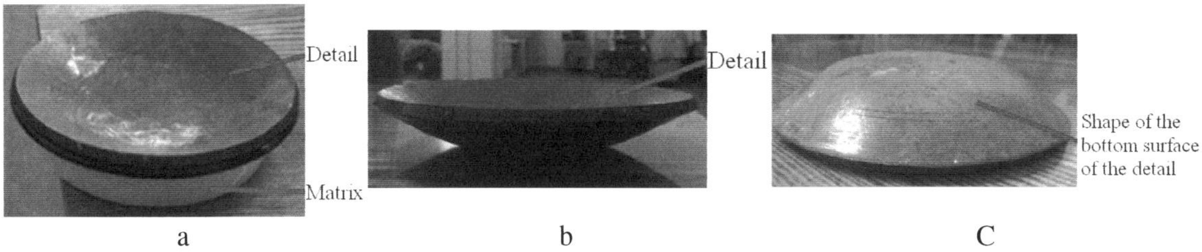

Figure 4. The obtained experimental sample: a) the part and the bottom base (matrix) of the package; b) general view of the part; c) the shape of the lower surface of the part.

3. Mathematical modeling of metal shaping and force parameters of the combined rolling and stamping process in conditions of realization on a laboratory mill duo 150

Mathematical modeling was carried out by the finite element method in the DEFORM 3D software package. The adaptation of the finite element mathematical model consisted in the scientifically justified formulation of the boundary and initial conditions of the problem being solved, as well as in the adoption of certain assumptions of the simulated process. The following boundary conditions were set: 1) the boundary surfaces of the working tool (matrix, punch, rolls); 2) the law of friction on contact with the working tool; 3) rheological model of the deformable material. As initial conditions for modeling, the following parameters were set: 1) diameters and angular speeds of rotation of working rolls; 2) the yield curve of the material (lead) of the billet (Fig. 5); 3) the temperature of the workpiece; 4) the initial dimensions of the workpiece; 5) the friction index m on contact with the tool; 6) the size and number of finite elements.

With three-dimensional modeling of the process, the following assumptions are accepted: 1) the deformable medium is viscoplastic; 2) matrix, punch are elastic, rolls are rigid (incompressible); 3) the contact friction stresses are proportional to the contact normal stresses.

The following quantities were determined when solving the three-dimensional problem: 1) the stress field; 2) the field of deformations; 3) the field of displacements.

We will consider further the results of a comparative analysis of the stress-strain state of metal during traditional stamping and the combined "rolling-stamping" process of "bottom" parts.

Determination of the stress-strain state (SSS) of a billet is not an end; SSS further allows calculating the deformation force, estimating the change in the thickness of the billet in the process of deformation, determining the plasticity resource of the material and predicting technological failures in the process of deformation.

The simulation results of the billet preparation with a curvilinear surface based on the traditional stamping process are shown in Figure 6 [1-3]. This figure shows the deformation fields after 20 and 32 steps.

The maximum strain rates reach values of 0.1. A general view of the modeled part and a graph of the change in punching force are shown in Figures 7 and 8. The maximum deformation force reaches 510 N. We have made a comparative analysis of the stress-strain state of the metal during the traditional stamping and the combined rolling and stamping" process of the "bottom" parts [4-6]. It should be noted that the use of high speed rolls and large single crimps in the rolling stand leads to the same defects as in the laboratory tests. In these cases, as has been already noted, the workpiece moves relative to the die and punch during the process.

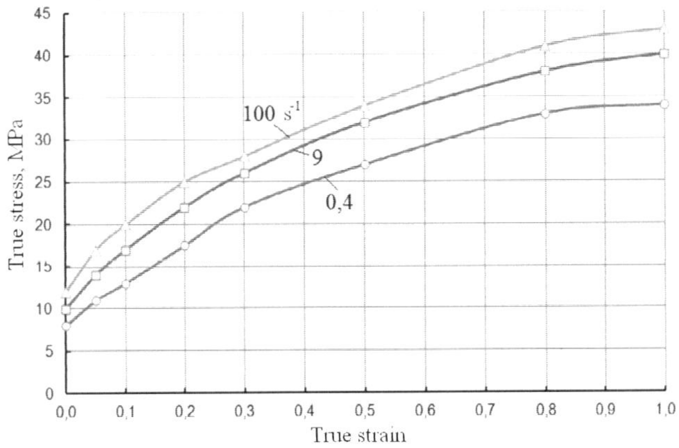

Figure 5. Stress of the material flow (lead) of the billet.

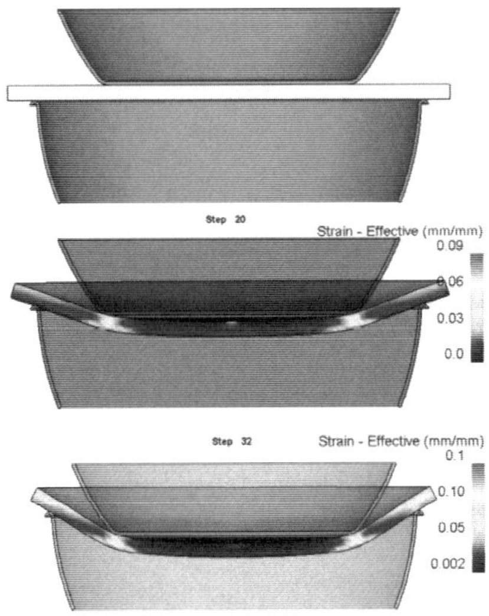

Figure 6. Simulation of the traditional stamping process: a) the initial state; b) after 20 steps (after 1.0 sec); c) at the end of the process (in 1.6 s)

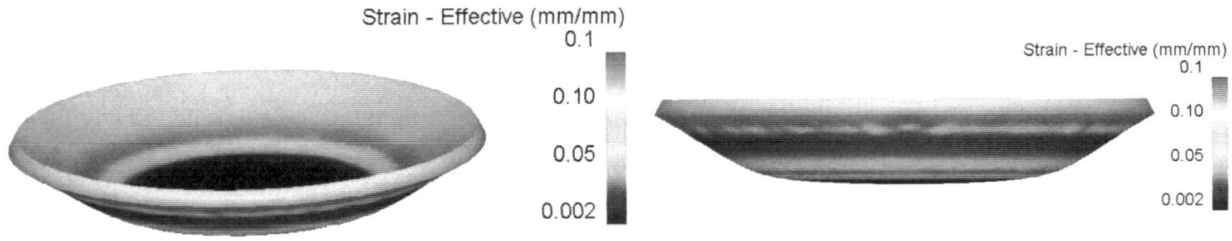

Figure 7. General view of the modeled bottom part and the strain intensity field (after laboratory punching).

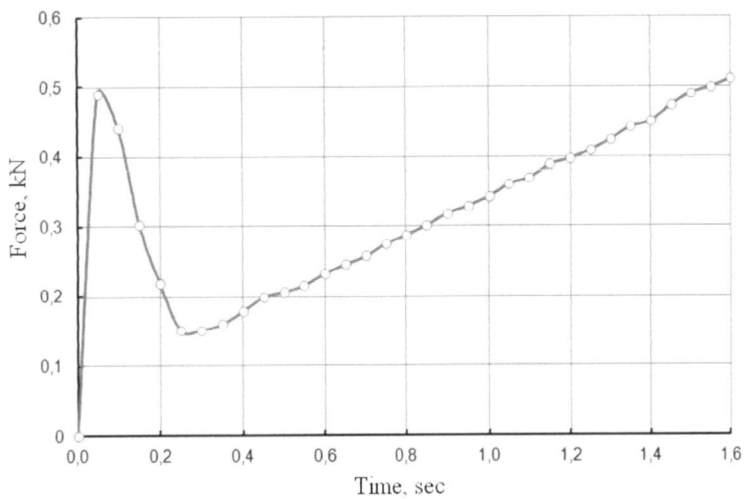

Figure 8. Graph of the change in the force of deformation in the production of a "bottom" type part by the method of traditional stamping.

The force of deformation at the combined process of rolling and stamping turned out to be almost two times less (Figure 9), in comparison with the usual stamping. At the same time, the processing time of the finished product increased from 1.6 seconds (when stamping) to 2.5 seconds (in a combined process).

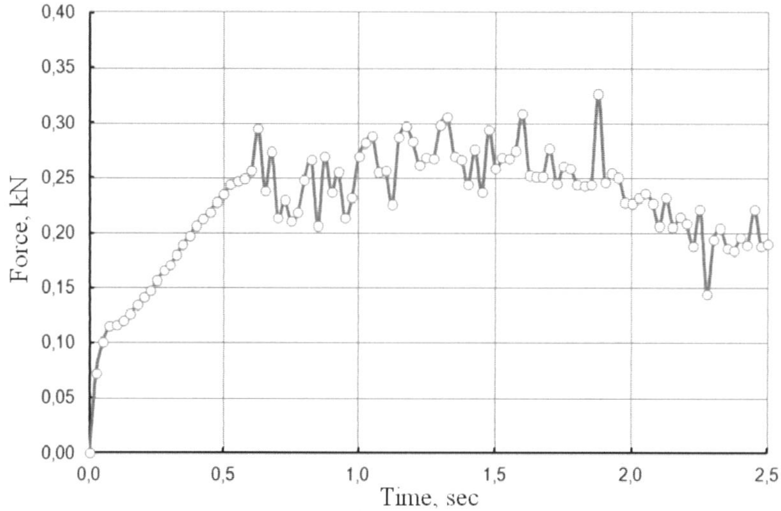

Figure 9. Graph of the change in the force of deformation in the production of a "bottom" type part by the "rolling-stamping" method.

4. Summary

An experimental study of the developed new technologies for obtaining large-sized parts with an elliptical surface has been carried out. The advantages and disadvantages are established, as well as the boundaries of effective applicability of new combined process. Verification of the results of computer simulation based on testing the technology in the conditions of a laboratory rolling mill was performed. The force of deformation at the combined process of rolling and stamping turned out to be almost two times less, in comparison with the usual stamping.

Acknowledgments

The reported study was supported by grant (program START 2014), research project No 190GS1/8745.

References

[1] Karbasian H. and Tekkaya A.E. 2010. A review on hot stamping. Journal of Materials Processing Technology. Vol. 210. P. 2103-2118

[2] Huanga Ming-dong, Wanga Baoyu, Zhoua Jing, Lib Xuetao. 2014. Investigation on Mechanical Properties Distribution on Hot Stamped Part Procedia Engineering. Vol. 81. P. 1774-1779

[3] Lei Chengxi, Cui Junjia, Xing Zhongwen, Fu Hongya, Zhao Hao. 2012. Investigation of Cooling Effect of Hot-stamping Dies by Numerical Simulation // Physics Procedia. Vol. 25. P. 118-124

[4] Pesin A., Drigun E, Pustovoytov D., Pesin I. 2016. Technology Development of Large-Size Bodies Manufacturing from Thick Plate Materials Based on Combined Methods of Deformation // Key Engineering Materials. 2016. Vol. 685, P. 375-379

[5] Pesin A., Drigun E., Pustovoytov D., Pesin I. 2016. Development of the Technology of Various Large Bodies Manufacturing Based on Combined Methods of Deformation // Key Engineering Materials. Vol. 716, P. 659-666

[6] Pesin A., Drigun E., Pustovoytov D., Pesin I. 2016. Finite Element Modeling of Combined Process of Plate Rolling and Stamping // MATEC Web of Conferences, vol. 80, 15008. P. 1-7

IOP Conf. Series: Materials Science and Engineering **293** (2017) 012007 doi:10.1088/1757-899X/293/1/012007

Finite element simulation and comparison of a shear strain and equivalent strain during ECAP and asymmetric rolling

A. Pesin, D. Pustovoytov, T. Shveyova, R. Vafin

Nosov Magnitogorsk State Technical University, 38, Lenin prospect, Magnitogorsk, 455000, Russia

Corresponding author – Alexander Pesin: pesin@bk.ru

Abstract. The level of a shear strain and equivalent strain plays a key role in terms of the possibility of using the asymmetric rolling process as a method of severe plastic deformation. Strain mode (pure shear or simple shear) can affect very strongly on the equivalent strain and the grain refinement of the material. This paper presents the results of FEM simulations and comparison of the equivalent strain in the aluminium alloy 5083 processed by a single-pass equal channel angular pressing (simple shear), symmetric rolling (pure shear) and asymmetric rolling (simultaneous pure and simple shear). The nonlinear effect of rolls speed ratio on the deformation characteristics during asymmetric rolling was found. Extremely high equivalent strain up to e=4.2 was reached during a single-pass asymmetric rolling. The influence of the shear strain on the level of equivalent strain is discussed. Finite element analysis of the deformation characteristics, presented in this study, can be used for optimization of the asymmetric rolling process as a method of severe plastic deformation.

1. Introduction

Severe plastic deformation (SPD) is a very effective way to improve mechanical properties of metallic materials due to creating an ultrafine grain structure [1]. Nowadays many variants of SPD methods are available. The best known and well-studied of them is equal channel angular pressing (ECAP) [2]. In many works [1-3] it was proved that grain refinement and formation of high angle boundaries during ECAP comes from large shear strain. The main limitation of the application of ECAP is the small size of the samples. Therefore ECAP has low potential for producing materials that may be used in a wide range of structural applications. Contrary to ECAP the asymmetric rolling (AR) can provide the possibility for overcoming the limitation of producing ultrafine grained materials with large dimensions due to its continuous feature [4-6]. Many works studied the shear strain and equivalent strain during ECAP and AR by using finite element method (FEM). Ji and Park analyzed various AR processes by the rigid-viscoplastic FEM [4]. The results of the numerical simulation demonstrated that shear strain is more severe in the material layers, where the circumferential speed or friction coefficient is greater. Kim et al. [7] analyzed the effect of speed ratio on the development of shear strain and texture during AR. FEM simulation results showed that the equivalent strain during AR increased with high-speed ratio. Influence of process parameters on distribution of shear strain through sheet thickness during AR was investigated by FEM in [8]. It was found that to obtain a high uniform shear strain through the sheet thickness the rolls speed ratio should be equal to thickness reduction per pass. The main disadvantage of AR is a formation of a vertical bending of the sheet. This creates significant difficulties in multi-pass AR. So it is very important to find the optimal process parameters, which can provide extremely high shear strain and equivalent strain up to e > 3.0 during a single-pass asymmetric rolling. The level of a shear strain and equivalent strain plays a key role in terms of the possibility of using the AR process as a method of SPD [9]. Two types of shear strain are known: pure shear and simple shear. During pure shear the axes of the strain ellipsoid do not rotate and the incremental and finite strain ellipsoids are coaxial. Simple shear involves rotation of the axes of the strain ellipsoid. Exactly rotation leads to grain refinement and formation of high angle boundaries. Simple shear (Fig. 1a) corresponds to ECAP. Pure shear (Fig. 1b) corresponds to symmetric rolling (SR). Simultaneous pure and simple shear (Fig.1c) corresponds to asymmetric rolling. Therefore, for better understanding of a maximum possible level of equivalent strain, generated during a single-pass SPD, it is necessary to compare different strain modes. The goal of this investigation is a numerical simulation and comparison of the equivalent strain in the sample processed by a single-pass ECAP (simple shear), symmetric rolling (pure shear) and asymmetric rolling (simultaneous pure and simple shear). The results of investigation can be used for design of technology of producing ultrafine grained materials.

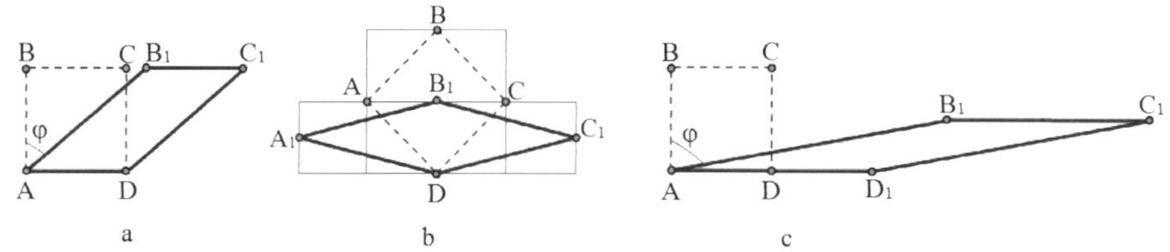

Figure 1. Schematic illustration of different strain modes:
simple shear (a), pure shear (b), simultaneous pure and simple shear (c).

2. Research method

Simulations of a single-pass ECAP, SR and AR were carried out using the commercial FEM code DEFORM 3D. The geometry model of ECAP is shown in Fig. 2a. For ECAP process, workpiece with dimensions of 8.0×8.0×40.0 mm^3 was used for processing. The workpiece was pressed through a die with two channels of equal cross section, intersecting at an angle 90° (channel angle). A Coulomb friction model was used. The friction coefficient at the die-workpiece interface was considered as 0.05, and the plunger moved with the velocity of 5.0 mm/sec during the process. The workpiece was meshed with ≈10000 brick elements.

Schematic illustration of the AR is shown in Fig. 2b. Workpiece with initial dimensions of 1.0×40.0×40.0 mm^3 was meshed with ≈40000 brick elements. AR was performed for a single pass with a thickness reduction 60%. Simulation was carried out at a constant angular speed of the top roll 0.4 rad/s. Angular speed of the bottom roll was reduced by 10...60%. The diameters of the rolls were 500 mm. A Coulomb friction model was used between rolls and workpiece. The friction coefficient on the contact surfaces was 0.4. The modeling of the all processes (ECAP, SR, AR) was performed at room temperature without taking into account the increment of the metal's temperature due to the thermal effect of deformation and friction. Aluminium alloy 5083 was chosen as a hardened rigid-plastic material for the workpiece in all calculation variants. The die, plunger and rolls were assumed as rigid. Symmetry plane was taken to improve computational efficiency. The stress-strain state of metal during a single-pass ECAP, SR and AR was compared and analyzed by FEM.

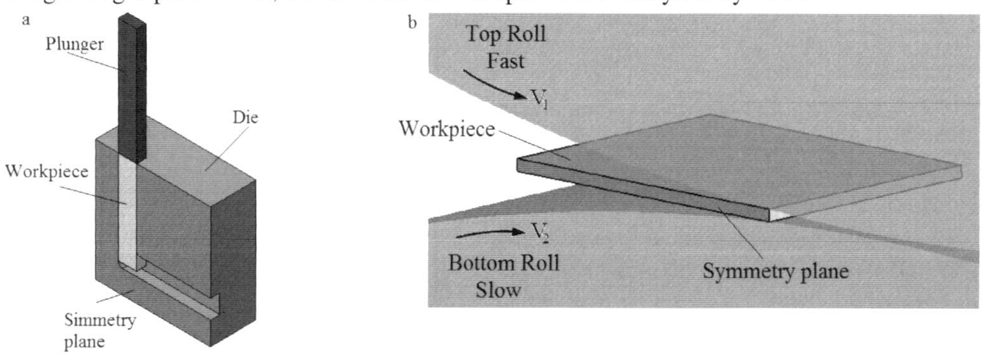

Figure 2. Geometry model of ECAP (a) and asymmetric rolling (b).

IOP Conf. Series: Materials Science and Engineering **293** (2017) 012007 doi:10.1088/1757-899X/293/1/012007

3. Simulation results and discussion

Simulation allowed to carry out a comparison between simple shear, pure shear and simultaneous pure and simple shear in order to highlight differences in the equivalent strain generated in the workpiece by a single pass ECAP, SR and AR. Changes in a single Lagrange's cell (along the centerline of the workpiece) before and after ECAP, SR and AR are shown in Figures 3, 5, 8 respectively. Single cell with initial shape of square becomes the parallelogram with no change in thickness after ECAP (Fig. 3). Shear strain during ECAP can be presented as tangent of shear angle φ. After a single pass ECAP shear angle is about 64 degrees. This leads to a high value of the equivalent strain, which is about e≈1.15 (Fig. 4). This fully corresponds to the analytical calculation of the equivalent strain during ECAP by the well-known formula:

$$e = \frac{tg\varphi}{\sqrt{3}} \quad (1)$$

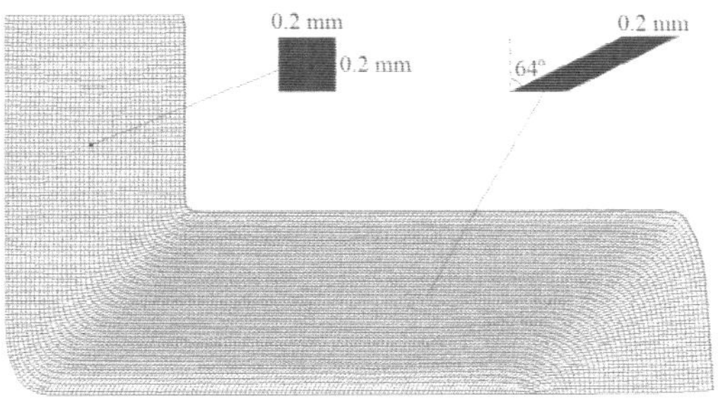

Figure 3. Change of Lagrange's cells during ECAP.

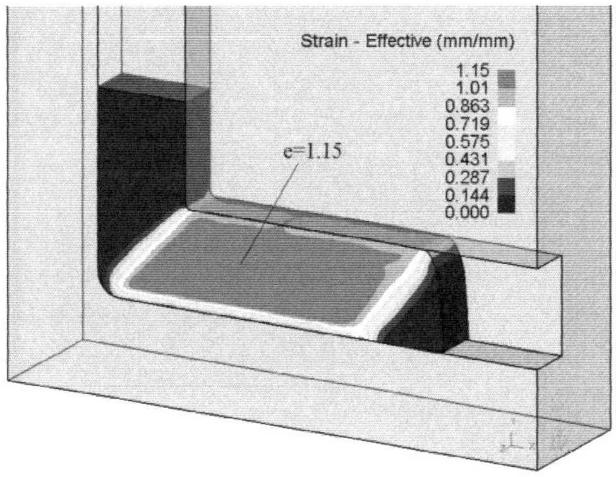

Figure 4. Distribution of the equivalent strain in the workpiece during ECAP.

Single cell with initial shape of square becomes the rectangle in the middle layer of the workpiece after symmetric rolling (Fig. 5). The thickness of the square is reduced in normal direction and the length is elongated toward rolling direction. It looks like the only deformation involved is compression and extension. However, if we examine the diagonals of the cell we see that there is indeed shear because the angle between the diagonals changes. This sort of shear is called pure shear. Single pass rolling with thickness reduction 60% leads to a high value of the equivalent strain, which is about e≈1.1 (Fig. 6). This corresponds to the analytical calculation of the equivalent strain during SR by the well-known formula:

$$e = \frac{2}{\sqrt{3}} \ln \frac{h_0}{h_1}, \quad (2)$$

where h_0 is an initial thickness; h_1 is a final thickness.

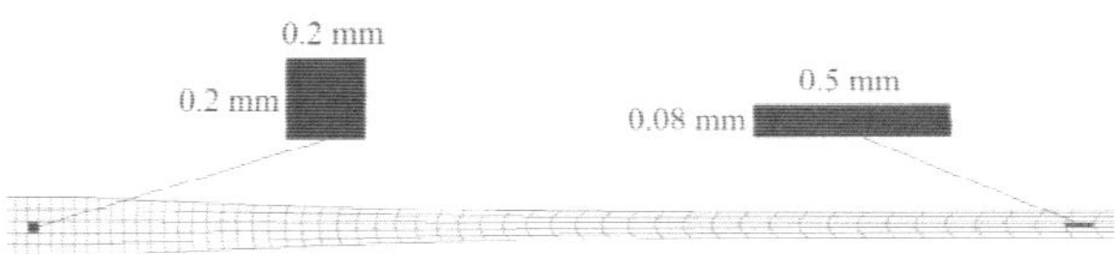

Figure 5. Change of Lagrange's cells during SR.

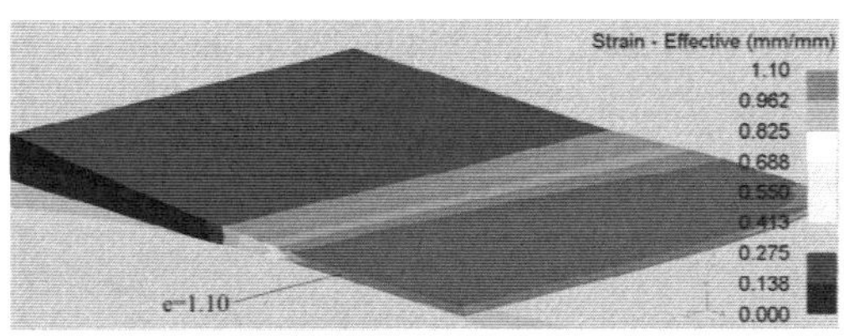

Figure 6. Distribution of the equivalent strain in the workpiece during SR.

Equivalent strain, generated in the workpiece during a single pass symmetric rolling by pure shear (with thickness reduction 60%), corresponds to equivalent strain during one pass ECAP by a simple shear. However, the key disadvantage of pure shear is the absence of a rotational strain mode, which is necessary for forming of high-angle grain boundaries.

The workpiece is deformed according to the scheme of simultaneous pure and simple shear during asymmetric rolling. Friction coefficient, the rolls speed ratio and the thickness reduction per pass are the main factors which provide synergistic effect on the amount of the shear strain and the equivalent strain during AR. Increasing the rolls speed ratio ΔV from 0 to 60% leads to serious increase of the equivalent strain (Fig. 7). The non-linear effect of the rolls speed ratio on the deformation characteristics during asymmetric rolling was found. The equivalent strain has the maximum value when the rolls speed ratio ΔV is optimal. Equivalent strain e≈4.2 when ΔV=57% (Fig.7).

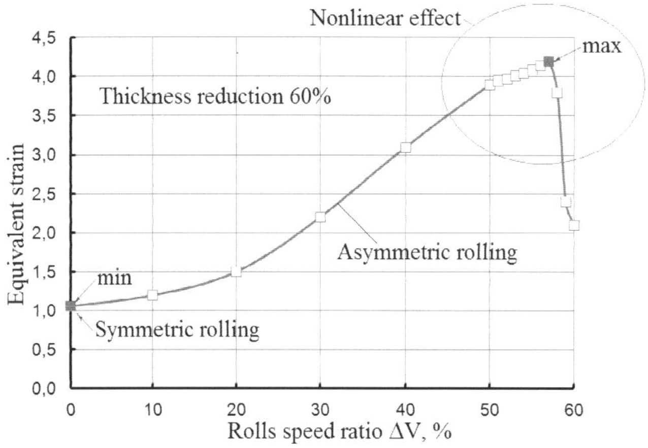

Figure 7. Influence of the rolls speed ratio ΔV on the equivalent strain of the workpiece during AR (initial thickness is 1.0 mm; thickness reduction is 60%; friction coefficient is 0.4).

Single cell with initial shape of square becomes the parallelogram with change in thickness after asymmetric rolling (Fig. 8). Shear strain also can be presented as tangent of shear angle φ. Single pass AR with thickness reduction 60% leads to a high value of the shear angle, which is about 83 degrees, and it leads to extremely high equivalent strain, which is about e≈4.2 (Fig. 9). This corresponds to the analytical calculation of the equivalent strain during AR by the formula [8]:

$$e = \frac{2}{\sqrt{3}}\sqrt{\left(\ln\frac{h_0}{h_1}\right)^2 + \frac{tg^2\varphi}{4}} \tag{3}$$

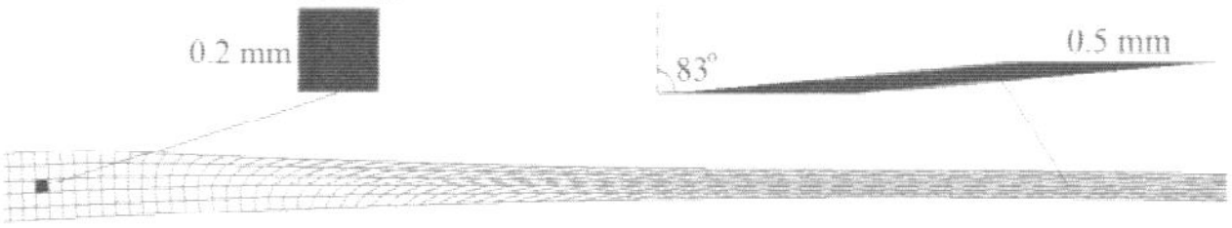

Figure 8. Change of Lagrange's cells during AR.

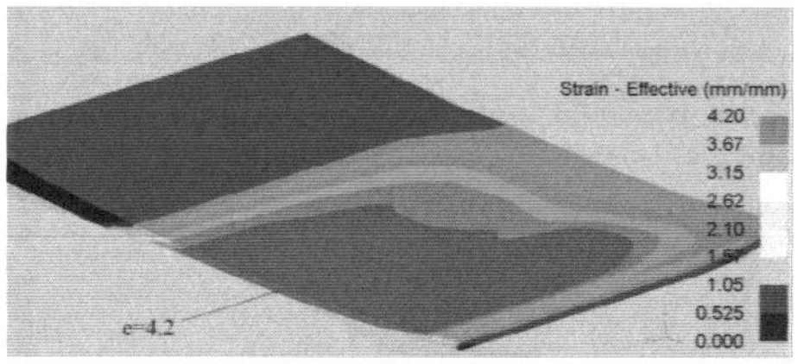

Figure 9. Distribution of the equivalent strain in the workpiece during AR.

If shear angle φ = 0, then formula (3) transforms into (2). This corresponds to a pure shear during symmetric rolling. If an initial thickness h_0 is equal to a final thickness h_1, then formula (3) transforms into (1). This corresponds to a simple shear during ECAP. The shape of Lagrange's cell after ECAP, SR and AR highlights the remarkable action of the shear strain developed by simple shear, pure shear, and simultaneous pure and simple shear.

Severe plastic deformation usually corresponds to the equivalent strain e > 3. This requires at least 3…4 passes of ECAP with a simple shear. A single pass asymmetric rolling with simultaneous pure and simple shear is preferable to ECAP with simple shear in developing of extremely high equivalent strain up to e=3…6, when shear angle φ is no less than 80° (Fig. 10).

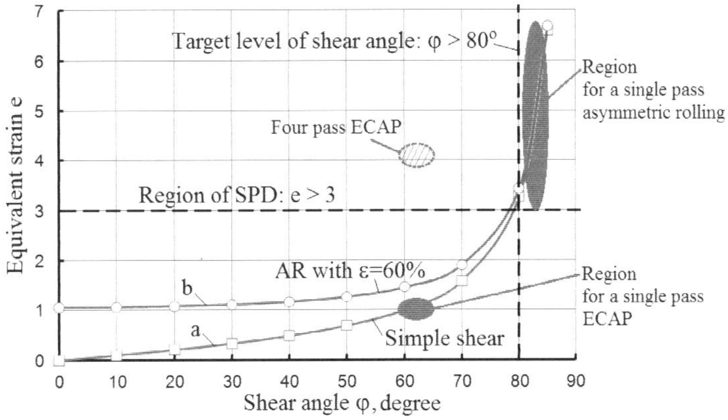

Figure 10. Influence of the shear angle φ on the equivalent strain.

4. Conclusions

Finite element simulation and comparison of the equivalent strain in the sample processed by a single-pass ECAP (simple shear), symmetric rolling (pure shear) and asymmetric rolling (simultaneous pure and simple shear), were performed. The non-linear effect of the rolls speed ratio on the deformation characteristics during asymmetric rolling was found. The equivalent strain has the maximum value when the rolls speed ratio ΔV is optimal. It was demonstrated that during a single-pass asymmetric rolling of a 1.0 mm thick aluminium alloy 5083 by work rolls of D = 500 mm with thickness reduction per pass $\varepsilon = 60\%$, friction coefficient 0.4 and rolls speed ratio $\Delta V = 57\%$ the extremely high shear strain and equivalent strain up to e=4.2 can be reached. A single pass asymmetric rolling with simultaneous pure and simple shear is preferable to ECAP with simple shear in developing of extremely high equivalent strain, when shear angle φ is no less than 80°. FE analysis of the deformation characteristics, presented in this study, can be used to optimize the asymmetric rolling process as SPD method to improve the microstructure and mechanical properties of metallic materials. Further experimental investigation of the microstructure evolution during asymmetric rolling is required.

Acknowledgments
The study was supported by a grant of the Russian Science Foundation (project No15-19-10030).

References

[1] Valiev RZ, Estrin Y, Horita Z, et al. 2006 Producing bulk ultrafine-grained materials by severe plastic deformation *JOM* **58** 33-39
[2] Valiev RZ, Langdon TG 2006 Principles of equal-channel angular pressing as a processing tool for grain refinement *Prog. Mater. Sci.* **51** 881-981
[3] Iwahashi Y, Horita Z, Nemoto M, Langdon TG 1998 The process of grain refinement in equal-channel angular pressing *Acta Mater.* **A382** 30-34
[4] Ji YH, Park JJ 2009 Development of severe plastic deformation by various asymmetric rolling processes *Mater. Sci. Eng. A* **499** 14-17
[5] Cui Q, Ohori K 2000 Grain refinement of high purity aluminum by asymmetric rolling *Materials Sc. and Technology* **16** 1095-1101
[6] Kim WJ, Lee KE, Choi SH 2009 Mechanical properties and microstructure of ultra fine-grained copper prepared by a high-speed-ratio differential speed rolling *Mater. Sci. Eng. A* **506** 71-79
[7] Kim WJ, Hwang BG, Lee MJ, Park YB 2011 Effect of speed-ratio on microstructure, and mechanical properties of Mg-3Al-1Zn alloy *Journal of Alloys and Compounds* **509** 8510-8517
[8] Pesin A, Pustovoytov D 2014 Influence of process parameters on distribution of shear strain through sheet thickness in asymmetric rolling *Key Engineering Materials* **622-623** 929-935
[9] Zuo F, Jiang J, Shan A 2008 Shear deformation and grain refinement in pure Al by asymmetric rolling *Transactions of Nonferrous Metals Society of China* **18** 774-777

IOP Conf. Series: Materials Science and Engineering **293** (2017) 012008 doi:10.1088/1757-899X/293/1/012008

Development of a multi-cycle shear-compression testing for the modeling of severe plastic deformation

A. Pesin, D. Pustovoytov, N. Lokotunina

Nosov Magnitogorsk State Technical University, 38, Lenin prospect, Magnitogorsk, 455000, Russia

Corresponding author – Alexander Pesin: pesin@bk.ru

Abstract. The mechanism of severe plastic deformation comes from very significant shear strain. Shear-compression testing of materials is complicated by the fact that a state of large equivalent strain with dominant shear strain is not easily achievable. This paper presents the novel technique of laboratory simulation of severe plastic deformation by multi-cycle shear-compression testing at room temperature with equivalent strain e=1...5. The specimen consisted of a parallelepiped having an inclined gauge section created by two diametrically opposed semi-circular slots which were machined at 45°. Height of the specimen was 50 mm, section dimensions were 25×25 mm, gauge thickness was 5.0 mm and gauge width was 6.0 mm. The specimen provided dominant shear strain in an inclined gauge-section. The level of shear strain and equivalent strain was controlled through adjustment of the height reduction of the specimen, load application direction and number of cycles of shear-compression. Aluminium alloy Al-6.2Mg-0.7Mn was used as a material for specimen. FE simulation and analysis of the stress-strain state were performed. The microstructure of the specimen after multi-cycle shear-compression testing with equivalent strain e=1...5 was examined by optical and scanning electron microscope.

1. Introduction

Severe plastic deformation (SPD) can be used to improve the microstructure and mechanical properties of metallic materials [1]. The mechanism of SPD comes from very significant shear strain. Laboratory simulation of stress-strain state and microstructure evolution, which are similar to that occurring during SPD, is very important for design of technologies of producing ultrafine grained materials. It can allow to consider a variety of technological variants and to find the best solution in relatively short time and with low costs. Shear-compression testing of materials is complicated by the fact that a state of large equivalent strain with dominant shear strain is not easily achievable. Problem of selection of shear testing and specimen geometry has been discussed in numerous publications. A shear-tensile specimen was developed by Hundy and Green [2]. Another example of such a specimen is the so-called "hat specimen" which was developed by Meyer and Manwaring [3], and used thereafter by many other researchers. Another variant is the double shear specimen used by Klepaczko [4]. Very simple specimen geometry (thin foil) has been used by Clifton and Klopp [5] in their pressure-shear plate impact experiments. Peirs et al. [6] introduced the eccentric notch shear specimen, which was designed for shear testing of sheet metals over a wide range of strain rates. Isakov, et al. [7] used modified specimen geometry, based on ASTM B831 [8], which presents specimen geometry intended to determine the ultimate shear strength of wrought or cast aluminium alloys. The specimen geometry consists of a planar specimen with a gauge section created by two opposing 45° slots through the specimen thickness. This sample geometry was modified for use in a tension Kolsky bar. It was shown that the specimen without the thickness reduction exhibits severe distortion outside of the intended gauge section. A shear-compression specimen for large strain testing of materials was developed by Rittel et al. [9]. The specimen consists of a cylinder having an inclined gauge section created by two diametrically opposed rectangular slots which are machined at 45° with respect to the longitudinal axis. The specimen was used in numerous investigations, including constitutive testing, texture evolution, and adiabatic shear banding. A miniature version of the shear-compression specimen was used for the characterization of nanomaterials. The modified shear-compression specimen was presented in [10]. In the modified specimen, the two diametrically machined gauges were semi-circular (instead of rectangular). This modification induces large strains on the mid-section of the gauge without the sharp edges and stress concentrations of the former rectangular gauges. The goal of this research is a development of a novel laboratory testing method which can be used for design of technology of SPD of fine grained aluminium alloys.

2. Research method

Simulation of a multi-cycle shear-compression testing was performed by using an AG IC Shimadzu universal testing machine equipped with a 300 kN load cell. Aluminium alloy Al-6.2Mg-0.7Mn with flow stress (Fig. 1) was used as a material for specimen (Table 1). The specimen (Fig. 2) consisted of a parallelepiped having an inclined gauge section created by two diametrically opposed semi-circular slots which were machined at 45°. Height of the specimen was 50 mm, section dimensions were 25×25 mm, gauge thickness was 5.0 mm and gauge width was 6.0 mm. The circular gauge had a radius of 3.0 mm. The specimen provided dominant shear strain in an inclined gauge-section. The specimen was compressed between two flat dies during shear-compression testing in accordance to the scheme (Fig. 3). Velocity of the top die displacement was 500.0 mm/min (8.33 mm/sec) in all variants. The level of shear strain and equivalent strain was controlled through adjustment of the height reduction of the specimen, load application direction and number of cycles of shear-compression. Each cycle of the shear-compression testing consisted of two steps. The first step included 10% of height reduction of specimen, after that specimen was rotated by 90º. The second step included reduction of the width from w to 25 mm (Fig. 3) for getting the quasi original shape of a parallelepiped. One cycle of shear-compression testing provided equivalent strain e≈1.0 in an inclined gauge-section. Multi-cycle shear-compression testing was performed at room temperature with equivalent strain up to e≈5. Numerical simulation and analysis of the stress-strain state during shear-compression testing was performed with the commercial finite element software DEFORM 3D. The specimen was meshed with 80000…100000 tetrahedral elements during simulation. A Coulomb friction model was used between dies and specimen. The dies were assumed as rigid. The specimen was an elastic-plastic. The friction coefficient on the contact surfaces was $\mu = 0.08$.

Table 1. The chemical composition of the aluminum alloy Al-6.2Mg-0.7Mn.

Si	Fe	Cu	Mn	Mg	Zn	Ti	Be	Al
0.14	0.25	0.03	0.73	6.2	0.08	0.04	0.0008	balance

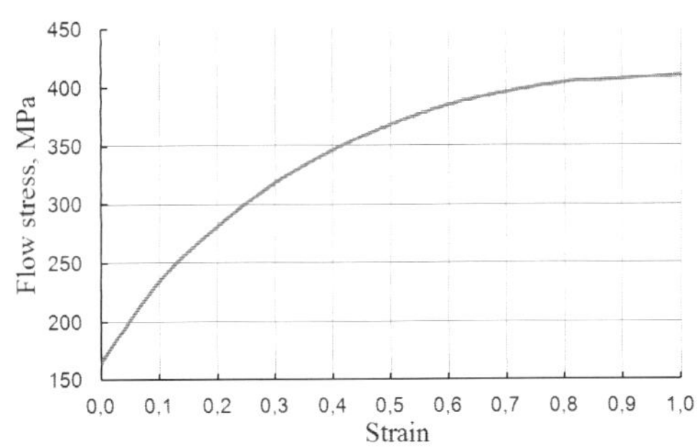

Figure 1. Stress-strain curve of Al-6.2Mg-0.7Mn.

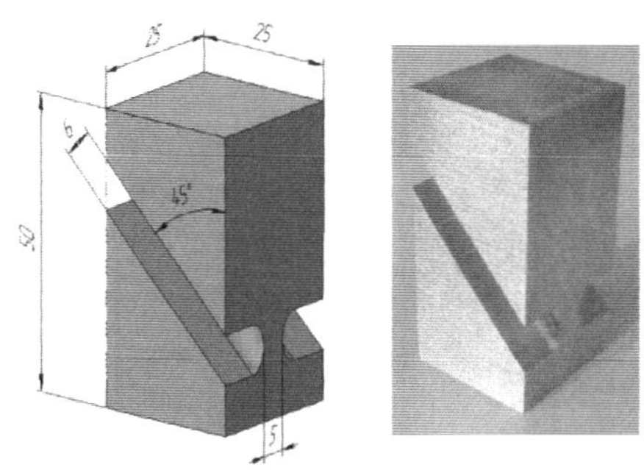

Figure 2. Specimen for shear-compression testing.

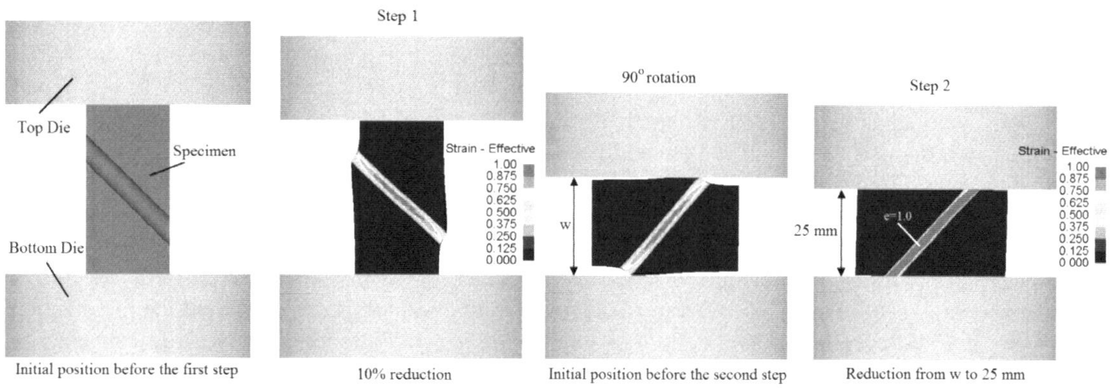

Figure 3. Scheme of the one cycle of the shear-compression testing.

3. Simulation results and discussion

The experimental shear-compression testing (Fig. 4) was computer controlled with registering the top die displacement and load. The comparison of experimental and numerical load-displacement curves during the first and the second steps of the first cycle of the shear-compression testing showed high accuracy (Fig. 5).

Figure 4. One cycle of the shear-compression testing: (a) initial position before the first step; (b) initial position before the second step.

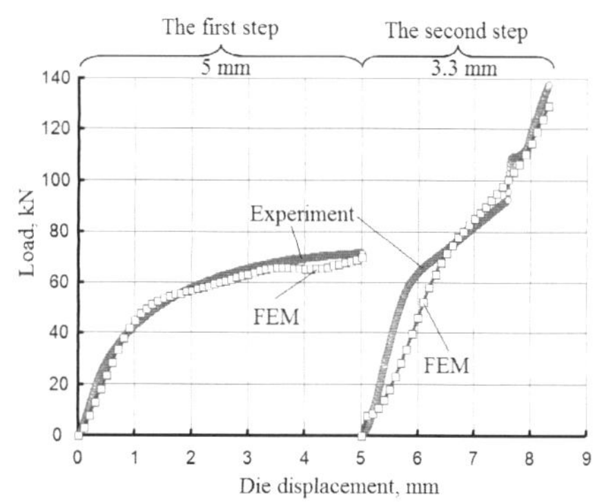

Figure 5. Load-displacement curves during the first and the second steps of the first cycle of shear-compression testing.

Asymmetry of the velocities of the metal flow is generated in gauge section during the each cycle of the shear-compression testing. Asymmetry of the velocities leads to high shear strain which is localized in gauge section. It should be noted that the height reduction of the specimen is the important factor that influence on the level of strain during shear-compression testing. With increasing the height reduction to 10% the effective strain is increased up to $e \approx 1.0$. So the level of strain can be controlled through adjustment of the height reduction of the specimen during shear-compression testing. However the height reduction of the specimen is limited by its destruction. The results of the stress-strain state of metal during shear-compression testing were analyzed. Distributions of the effective stress, strain and strain rate along the mid-section of the specimen are shown in Fig. 6. The uniform distributions can clearly be observed. High level of strain ($e \approx 1.0$) is localized in gauge section and it is typical for large plastic deformation.

The microstructure in gauge section of the specimens after 1...5 cycles of the shear-compression testing was examined by optical and scanning electron microscope (Fig. 7). Significant grain refinement was observed in Al-6.2Mg-0.7Mn after 5 cycles ($e \approx 5$) of the shear-compression testing.

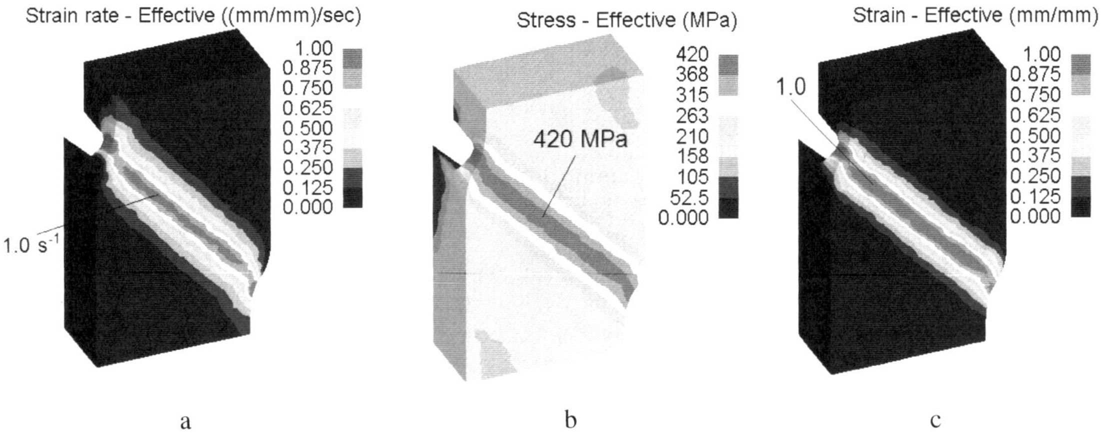

Figure 6. Fields of the effective strain rate (a), effective stress (b) and effective strain (c) during shear-compression testing.

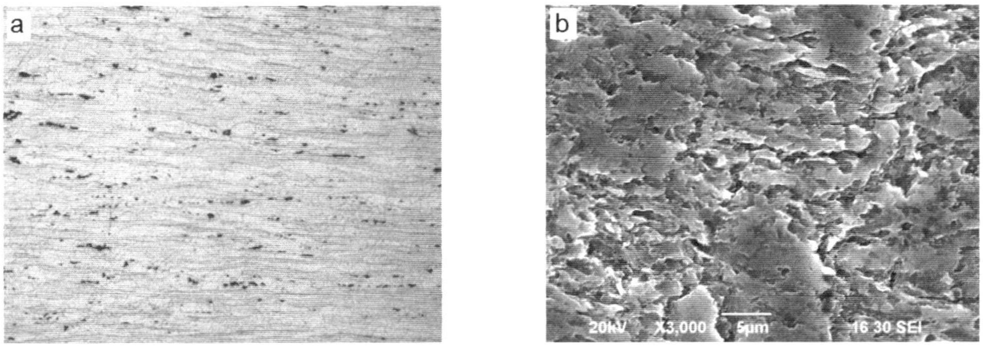

Figure 7. Microstructure after one (a) (×100) and five cycles (b) (×3000) of the shear-compression testing.

4. Conclusions

A novel technique of laboratory simulation of SPD by multi-cycle shear-compression testing was presented. The specimen consisted of a parallelepiped having an inclined gauge section created by two diametrically opposed semi-circular slots which were machined at 45°. Height of the specimen was 50 mm, section dimensions were 25×25 mm, gauge thickness was 5.0 mm and gauge width was 6.0 mm. The circular gauge had a radius of 3.0 mm. The specimen provided dominant shear strain in an inclined gauge-section. The level of shear strain and equivalent strain was controlled through adjustment of the height reduction of the specimen, load application direction and number of cycles of shear-compression. Aluminium alloy Al-6.2Mg-0.7Mn was used as a material for specimen. FE simulation and analysis of the stress-strain state were performed. Asymmetry of the velocities of the metal flow leads to high shear strain in gauge section of the specimen during the shear-compression testing. Uniform distributions of the effective stress and strain were shown. Significant grain refinement was observed in Al-6.2Mg-0.7Mn after 5 cycles ($e \approx 5$) of the shear-compression testing. The results of investigation can be used for design of technology of producing of fine grained aluminium alloys.

Acknowledgments
The study was supported by a grant of the Russian Science Foundation (project No15-19-10030).

References

[1] Valiev RZ, Estrin Y, Horita Z, et al. 2006 Producing bulk ultrafine-grained materials by severe plastic deformation *JOM* **58** 33-39

[2] Hundy BB, Green AP 1954 A determination of plastic stress-strain relations *Journal of the Mechanics and Physics of Solids* **3** 16-21

[3] Meyer LW, Manwaring S 1986 Metallurgical applications of shock-wave and high-strain-rate phenomena *Marcel Dekker*, New York, 1986

[4] Klepaczko JR 1994 An experimental technique for shear testing at high and very high strain rates *International Journal of Impact Engineering* **46** 25-39

[5] Clifton RJ, Klopp RW 1985 Pressure shear plate impact testing, *Metals Handbook* **8** 230-239

[6] Peirs J, Verleysen P, Degrieck J 2011 Novel technique for static and dynamic shear testing of Ti6Al4V sheet *Experimental Mechanics* **52** 729-741

[7] Isakov M, Seidt J, Ostman K, Gilat A, Kuokkala VT 2011 Characterization of a ferritic stainless sheet steel in simple shear and uniaxial tension at different strain rates *Proceedings of the ASME 2011 International Mechanical Engineering Congress and Exposition*

[8] ASTM. B831: Standard test method for shear testing of thin aluminum alloy products, ASTM International, 2005

[9] Rittel D, Lee S, Ravichandran G 2002 A shear-compression specimen for large strain testing *Experimental Mechanics* **42** 58-64

[10] Dorogoy A, Rittel D, Godinger A 2015 Modification of the shear-compression specimen for large strain testing *Experimental Mechanics* **55** 1627-1639

IOP Conf. Series: Materials Science and Engineering **293** (2017) 012009 doi:10.1088/1757-899X/293/1/012009

Effects of heat treatment on microstructure and mechanical behaviour of additive manufactured porous Ti6Al4V

S M Ahmadi[1], R K Ashok Kumar Jain[1], A A Zadpoor[1], C Ayas[1], V A Popovich[1].

1. Delft University of Technology, Faculty of Mechanical, Maritime, and Materials Engineering, Mekelweg 2, 2628 CD, Delft, The Netherlands, s.m.ahmadi@tudelft.nl

Abstract. Titanium and its alloys such as Ti6Al4V play a major role in the medical industry as bone implants. Nowadays, by the aid of additive manufacturing (AM), it is possible to manufacture porous complex structures which mimic human bone. However, AM parts are near net shape and post processing may be needed to improve their mechanical properties. For instance, AM Ti6Al4V samples may be brittle and incapable of withstanding dynamic mechanical loads due to their martensitic microstructure. The aim of this study was to apply two different heat treatment regimes (below and above β-transus) to investigate their effects on the microstructure and mechanical properties of porous Ti6Al4V specimens. After heat treatment, fine acicular α′ martensitic microstructure was transformed to a mixture of α and β phases. The ductility of the heat-treated specimens, as well as some mechanical properties such as hardness, plateau stress, and first maximum stress changed while the density and elastic gradient of the porous structure remained unchanged.

1. Introduction

Numerous products have been influenced by advent of additive manufacturing (AM). Complex designs and structures with a wide range of materials and sizes, which would have previously been just a fantasy, are nowadays available in the market, all thanks to AM [1]. AM itself has been evolved over time, changing from a simple method of prototyping to manufacturing high precision products in high-tech industries [2]. A wide range of materials are developed for AM from polymers to metals among which titanium has gained a remarkable position.

Given their outstanding mechanical properties, titanium and its alloys are widely used in various industries including the aerospace, automotive, and medical industries [3, 4]. In particular, titanium is a preferred material for fabrication of orthopaedic implants due to its three main advantages: lightness, biocompatibility, and strength [5]. Solid titanium implants remained unrivalled for decades, until the introduction of the new generation of porous titanium made by AM. AM porous implants have two substantial benefits: space for bone ingrowth and mimicking the mechanical properties of bone [6, 7]. Tailored AM porous titanium meets almost all the biomedical design requirements. However, it requires post-processing, since AM delivers near net shape and not the final product. So far, only a few studies have focused on the effects of different heat treatments on such porous structures. The aim of the current study is to investigate the effects of heat treatment on the microstructure and mechanical properties of AM Ti6Al4V porous structures. Two different heat treatment regimens (below and above β-transus) were selected and applied to the specimens. The resulting static mechanical properties such as the elastic gradient and plateau stress as well as the microstructure and hardness profile of the heat treated samples were compared with the as-processed material.

2. Materials and methods

Porous Ti6Al4V specimens with dimension of Φ15 mm x 20 mm were manufactured using a commercial selective laser melting (SLM) machine (SLM125, Realizer GmbH, Borchen, Germany) with the highest laser power of 400 W (IPG Photonics Corporation, Oxford, USA) and wavelength range of 1070 ± 10 nm. The specimens were designed using the regular diamond unit cell [8] with a unit cell size of 1 mm and an average relative density of 30.25 % (±1.5), which was measured by the dry weighting technique.

Three different groups were defined:

1. *As-processed* samples which are considered as the reference material.
2. Samples heat treated at 800 °C, further denoted as *T800*. For this group, the *as-processed* samples were first stress relieved at 600 °C, followed by tempering at 800 °C for 4 hours.
3. Samples heat treated at 1050 °C, further denoted as *T1050*. For this group, the *as-processed* samples were first stress relieved at 600 °C and then heat treated at 1050 °C for two hours.

The heat treatments were performed under an inert atmosphere (Argon) with a heating rate of 10° C/min, after which the furnace was allowed to cool down to the room temperature.

The mechanical properties of Ti6Al4V samples were determined using quasi-static compression tests with a static test machine (INSTRON 5985, 100 kN load cell). A constant deformation rate of 1.8 mm/min was applied to three specimens per group. The mechanical properties such as the elastic gradient, plateau stress, and first maximum stress were calculated in accordance with NEN ISO 13314:2011 [9]. Structure density, which is the ratio of the density of each strut to density of the same strut if it was 100 % solid, was calculated by measuring the internal pore area of the strut and deduct the number from the total surface area.

The micro-hardness measurements were carried out using an automated Vickers hardness machine, Dura Scan (Steurs) with a load of 0.3 Kgf. The hardness measurements were carried out along the cross section of the sample (XY plane) to analyze the influence subsequent layer deposition during SLM along the building direction.

The specimens were first ground (SiC 320 and MD Largo 9 μ m) and polished (MD Chem with OPS solution 0.04 μ m). Following which the samples were etched by immersion for 12-15 seconds in a mixture of 50 ml distilled water, 25 ml HNO_3 and 5 ml HF solution. The samples were observed using an optical microscope Keyence VHX5000.

X-Ray diffraction (XRD) patterns were obtained using Bruker D8 Advance diffractometer Bragg-Brentano geometry with graphite monochromator and Vantec position sensitive detector and CoKα radiation with the following measurement method: Diffraction patterns were scanned with 2θ range of 10 ° to 110 °, step size 0.034 ° 2θ and counting time per step 8 s.

3. Results and discussion

The compressive stress strain curves of *as-processed* and heat treated samples are illustrated in figure 1. From these graphs, static mechanical properties were extracted and presented in table 1. These results indicate that *as-processed* and below β-transus (*T800*) samples show similar mechanical behaviour. It should however be noted, that there is a slight difference in trend of the stress-strain curve after 0.5 compressive strain (figure 1). The reason behind this trend could be the effect of non-optimised processing parameters, which is out of the scope of this paper and will be presented in our follow up publication. Structure density of three groups were in the same range, indicating that another post treatment methods such as hot isostatic pressing (HIP) should be applied in order to reduce process induced porosity and defects [10].

In contrast with the first two groups, the specimens heat treated above the β-transus (*T1050*) showed different mechanical behaviour. Although the elastic gradient is in the same range as for the two other groups, however, the first maximum stress shows significant decrease while the plateau stress substantial increase.

The other main difference is the energy absorption (the energy required for deforming a specimen up to 50 % strain, i.e. area under the strain-stress curve) which is higher for *T1050* specimens. The obtained results indicate a ductile behaviour of *T1050* samples, which can be further explained by looking at their microstructure.

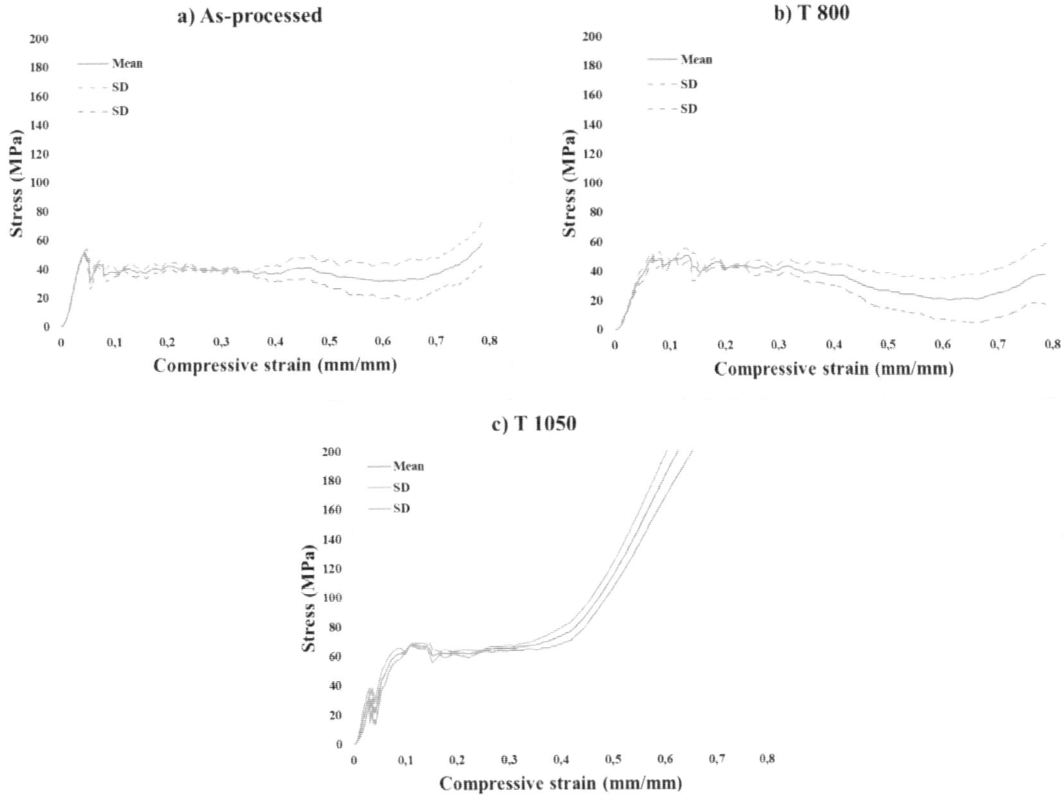

Figure 1. Compression stress-strain curve of a) *as-processed*, b) *T800* and, c) *T1050*

A fully fine acicular α´ martensitic microstructure was observed in *as-processed* samples (figure 2 a) and XRD data (figure 3) did not reveal any β phase. After stress relieving at 600 °C followed by heat treatment at 800 °C (*T800*) the martensitic fine α´ structure transformed to a combination of α and β phases. As shown in table 1, the β-fraction for *T800* samples is about 10 %. Comparing figure 2 a and figure 2 b, it is observed that α plates are coarser changing from less than 1 μm width to about 4 μm. Heat treatment at 1050 °C (*T1050*) results in a larger amount of β-fraction (table 1), which is in a good agreement with the bulk AM samples [11]. According to literature and phase diagrams [12], only β phase is expected when heating up the specimens above 995 °C (β transus), However, once specimens are furnace cooled, a mixture of lamellar α+β appears (Figure 2.c). It should also be noted that a further coarsening of α plates is observed for this heat treatment regime.

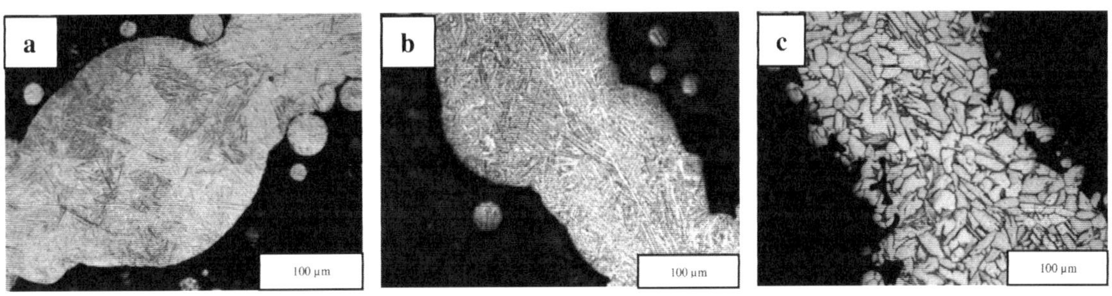

Figure 2. Microstructure of a) *as-processed*, b) *T800* and c) *T1050*

XRD data show significant peak shifting of *as-processed* and heat treated samples (figure 3). Peak shifting in the XRD data for the heat treated samples can be attributed to the formation of the equilibrium phases consisting of a two phase lamellar (α+β) structure. The transformation occurs from the non-equilibrium α´ phase in the *as-processed* condition (due to the high cooling rate involved in the SLM process) to the equilibrium α phase for the heat treated samples [11]. This transformation can explain the decrease in the width and increase in the intensity of the XRD peaks. It should also be noted, that peak shifting might also be effected by residual stress, which is rather unlikely since as-processed and stress relived (T800) samples do not show any significant differences (see an enlarged insert in Figure 3). For the two heat treated *T800* and *T1050* samples, a peak shift was observed on the specimens surface (Figure 3). However, there were no significant peak shifting observed when measurements were performed in cross section (in core) of the heat treated sample (Figure 3). This could be due to the high temperature involved in the heat treatment resulting in a possible solute redistribution from the surface which causes a change in the lattice parameters for the two phases and hence the peak shift for the 1050 °C heat treated sample at the surface to lower 2θ angles.
Another important observation from Figure 3 is the reduction in the peaks width of the heat treated samples when compared to the as processed material, which is most likely attributed to the increased crystallinity size observed in Figure 2 [13].

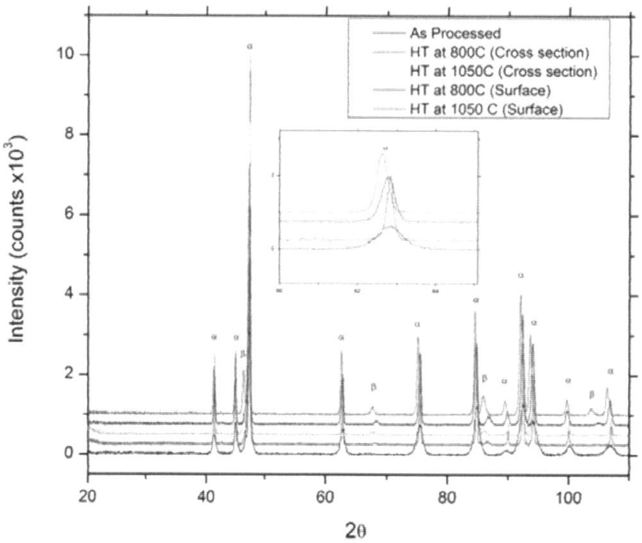

Figure 3. XRD measurement of *as-processed*, *T800* and *T1050*.

Finally, the results of hardness measurement showed reduction from 412 (±23) of *as-processed* samples to 353 (±34) for *T1050*. Hardness reduction can be explained by microstructural evolution from needle-shape α´ martensitic structure to a mixture of α+β and by increment of β phase, shown in table 1).

Table 1. Mechanical properties

	Elastic gradient (SD) GPa	Hardness (SD) $HV_{0.3}$	Platea stress (SD) MPa	Max stress (SD) MPa	β fraction (SD) %	Structure density (SD) %
As-processed	1.65 (0.046)	412.24 (23)	38.53 (0.63)	52.35 (1.91)	0	98.96 (0.74)
T800	1.60 (0.058)	388.15 (45)	42.44 (2.24)	52.50 (1.17)	9.6 (0.6)	98.15 (0.47)
T1050	1.49 (0.22)	353.25 (34)	66.34 (2.63)	38.41 (1.85)	20.4 (0.5)	98.61 (0.58)

4. Conclusion

The effect of different heat treatments (below and above β-transus) on microstructure and mechanical properties of porous additive manufactured Ti6Al4V structures was studied. It was observed that heat treatment substantially changes the microstructure of *as-processed* Ti6Al4V samples. XRD data showed peak shifting for both heat treatment regimes. However, the only considerable mechanical properties alteration was for plateau stress of *T1050* (heat treated above β-transus). Hardness profile showed reduction by incensement of temperature. The selected post treatment regimes produced a mixture of α+β phases and as previously observed [14] α+β interface is an effective barrier of crack propagation, which makes these structures attractive for load bearing applications. It is suggested to carry out further fatigue tests in order to investigate the impact of microstructure modification on dynamic mechanical properties of porous Ti6Al4V structures. Also, it is concluded that structure density of AM structures does not improve by heat treatment and in order to solve this problem other post treatments method such as hot isostatic pressing (HIP) must be considered.

References

[1] Vaezi M, Chianrabutra S, Mellor B, Yang S. Multiple material additive manufacturing–Part 1: a review: this review paper covers a decade of research on multiple material additive manufacturing technologies which can produce complex geometry parts with different materials. Virtual and Physical Prototyping 2013;8:19-50.
[2] Gibson I, Rosen D, Stucker B. Additive manufacturing technologies: 3D printing, rapid prototyping, and direct digital manufacturing: Springer; 2014.
[3] Gurrappa I. Characterization of titanium alloy Ti-6Al-4V for chemical, marine and industrial applications. Materials Characterization 2003;51:131-9.
[4] Cui C, Hu B, Zhao L, Liu S. Titanium alloy production technology, market prospects and industry development. Materials & Design 2011;32:1684-91.
[5] Elias C, Lima J, Valiev R, Meyers M. Biomedical applications of titanium and its alloys. Jom 2008;60:46-9.
[6] Zadpoor AA, Malda J. Additive manufacturing of biomaterials, tissues, and organs. Springer; 2017.
[7] Ahmadi SM, Yavari SA, Wauthle R, Pouran B, Schrooten J, Weinans H, et al. Additively manufactured open-cell porous biomaterials made from six different space-filling unit cells: The mechanical and morphological properties. Materials 2015;8:1871-96.
[8] Ahmadi S, Campoli G, Yavari SA, Sajadi B, Wauthlé R, Schrooten J, et al. Mechanical behavior of regular open-cell porous biomaterials made of diamond lattice unit cells. Journal of the mechanical behavior of biomedical materials 2014;34:106-15.
[9] 13314:2011 I. Mechanical testing of metals -- Ductility testing -- Compression test for porous and cellular metals. 2011.
[10] Tammas-Williams S, Withers PJ, Todd I, Prangnell PB. The Effectiveness of Hot Isostatic Pressing for Closing Porosity in Titanium Parts Manufactured by Selective Electron Beam Melting. Metallurgical and Materials Transactions A 2016;47:1939-46.
[11] Vrancken B, Thijs L, Kruth J-P, Van Humbeeck J. Heat treatment of Ti6Al4V produced by Selective Laser Melting: Microstructure and mechanical properties. Journal of Alloys and Compounds 2012;541:177-85.
[12] Weiss I, Froes FH, Eylon D, Welsch GE. Modification of alpha morphology in Ti-6Al-4V by thermomechanical processing. Metallurgical Transactions A 1986;17:1935-47.
[13] Popovich VA, Borisov EV, Popovich AA, Sufiiarov VS, Masaylo DV, Alzina L. Impact of heat treatment on mechanical behaviour of Inconel 718 processed with tailored microstructure by selective laser melting. Materials & Design 2017;131:12-22.
[14] Patil S, Kekade S, Phapale K, Jadhav S, Powar A, Supare A, et al. Effect of α and β Phase Volume Fraction on Machining Characteristics of Titanium Alloy Ti6Al4V. Procedia Manufacturing 2016;6:63-70.

Study of pyrolysed acid and based treated coconut coir as green photocatalyst substrate

Nilofar Asim [1*], Zeynab Emdadi [2], N A Abdullah [1], Masita Mohammad [1], Marzieh Badiei [3], Kamaruzzaman Sopian [1]

[1]Solar Energy Research Institute, Universiti Kebangsaan Malaysia, 43600 Bangi, Selangor, Malaysia
[2]Catalysts and Organic Synthesis Research Laboratory, Department of Chemistry, Iran University of Science and Technology, P.O. Box 1684613114, Tehran, Iran
[3] Agriculture & Natural Resource Research Centre of Khorasan Razavi , University of Applied Science & Technology,Mashahd,Iran

*nilofarasim@ukm.edu.my; nilofarasim@gmail.com

Abstract. This study investigates the possible contribution to sustainable development by utilizing agriculture waste materials to prepare a substrate for photo-catalysis application. The photocatalytic performance of impregnated TiO_2 on acid and base- treated coconut coir (CC) and their pyrolysed form have been studied. The photocatalytic performance of impregnated TiO2 on acid treated CC improved compared to bare TiO2. However, the pyrolysed samples showed higher thermal stability and porosity compared to only treated CC, their catalytic performance was decreased. It seems that impregnated TiO2 undergo interaction with treated CC during pyrolysis. More investigations to reveal exact reason of this behavior is in progress.

1. Introduction

Although theoretically all carbonaceous material with a high percentage of carbon in their composition could produce activated carbon (AC), only coal and lignocellulosic materials have been used for commercial production of activated carbon (ACs). Many researchers are investigating utilizing of agricultural by-products and residual wastes as resource materials for preparing the AC [1, 2]. It is well known that prepared activated carbon composition, structure, qualities and characteristics determine by the physical and chemical properties of the starting materials, the activating agent, and the preparation procedure [3]. Our previous studies on chemical treatment of agricultural wastes have been revealed that desirable changes in their structure are occurred which lead to higher water absorbency because of increasing of OH groups on their surfaces [4]. With this in mind, we decided to explore the possibility of using physical and chemical activation and treatment to prepare green substrate from the agricultural waste for catalytic application. Coconut coir is a natural fiber obtained from the coconut husk, a waste biomass readily available and a cheap material resulting in from the coconut fiber industry. Its fibers are bound together through lignin, cellulose and hemi-cellulose natural polymers [5]. In this study, nano TiO_2 has impregnated on the surface of coconut coir when it is chemically treated. Later, the prepared samples are pyrolysed and the photocatalytic behavior of them is studied.

2. Materials and Catalysts Preparation

After washing the coconut coir(CC) with distilled water and drying in the oven at 50 °C, they it was soaked in NaOH (QReC) 1.6M and nanoTiO_2(Aldrich, 21 nm) (5% W/W CC) and mixed in room temperature for 60 min. Later, it was washed with plenty of water till neutralized and dried at room temperature (RT) (hereafter denoted as sample CC(NaOH)). The dried materials were pyrolysed at 450°C for 1 hour under N_2 gas flow (heating rate 10 °C/min) thereafter prepared sample was utilized as catalyst in photocatalytical performance evaluation with methylene blue under UV light. For second try, the washed and dried coconut coir soaked with the H_2SO_4 (R&M Chemicals) solution 0.4 M and nano TiO_2 (5% W/W CC) and mixed at room temperature (RT) for 40 min. After washing with plenty of water till neutralized, it was let to dry at RT (hereafter denoted as sample CC (H_2SO_4)), the dried sample was pyrolysed at 450°C for 1 hour under N_2 gas flow (heating rate 10 °C/min).

3. Results and Discussion:

Figure. 1 shows the scanning electron micrographs of the prepared samples' surface (Carl Zeiss Merlin Compact). As shown in pictures, the Pyrolysis of both acid and base treated CC , makes porous honey comb shape in CC.

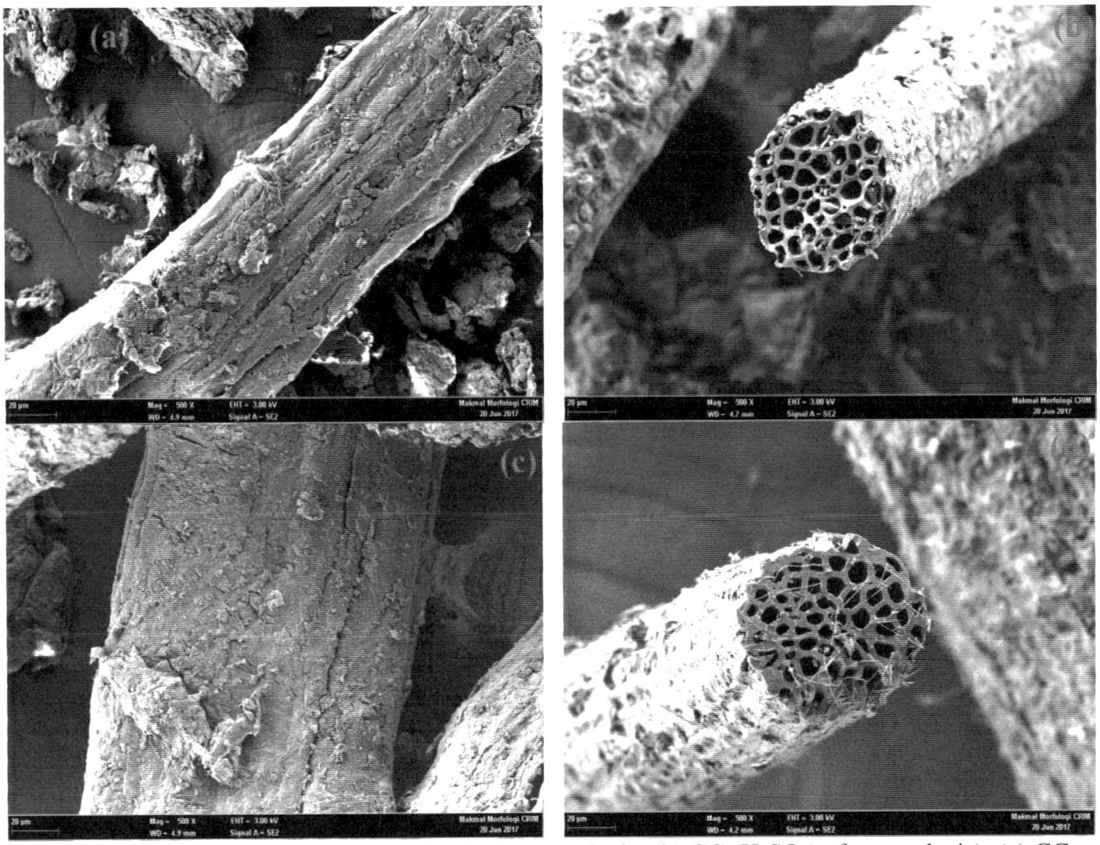

Figure.1 SEM images (a) CC (H_2SO_4) before pyrolysis, (b) CC (H_2SO_4) after pyrolysis), (c) CC (NaOH) before pyrolysis and (d) CC (NaOH) after pyrolysis, respectively.

TGA analysis (Figure 2) of prepared samples revealed that thermal stability of pyrolysed CC in both acid and base treated forms are much higher than CC without pyrolysis.

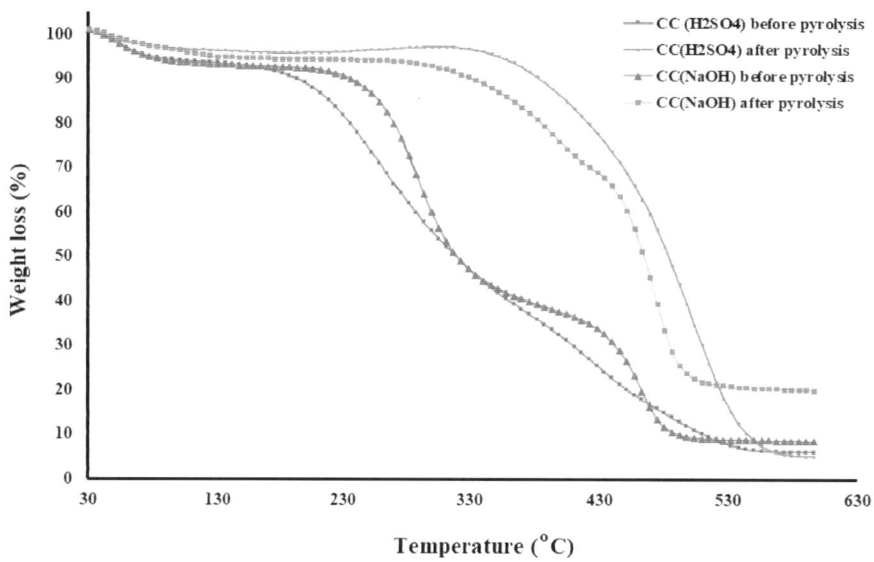

Figure2. TGA results for prepared samples

Crystallinity analysis of the prepared CC samples were conducted with a Bruker DB-Advance X-ray diffractometer (Figure 3). The peaks of impregnated TiO_2 nanoparticles could be seen in the XRD patterns of CC samples which treated with acid and base before pyrolysis while surprisingly, the peaks become wider and weaker after pyrolysis. This indicated that TiO_2 nanoparticles have the intercalation with treated CC during pyrolysis which caused the TiO_2 lattice structure to collapse and the intercalated product is relatively poorly crystalline.

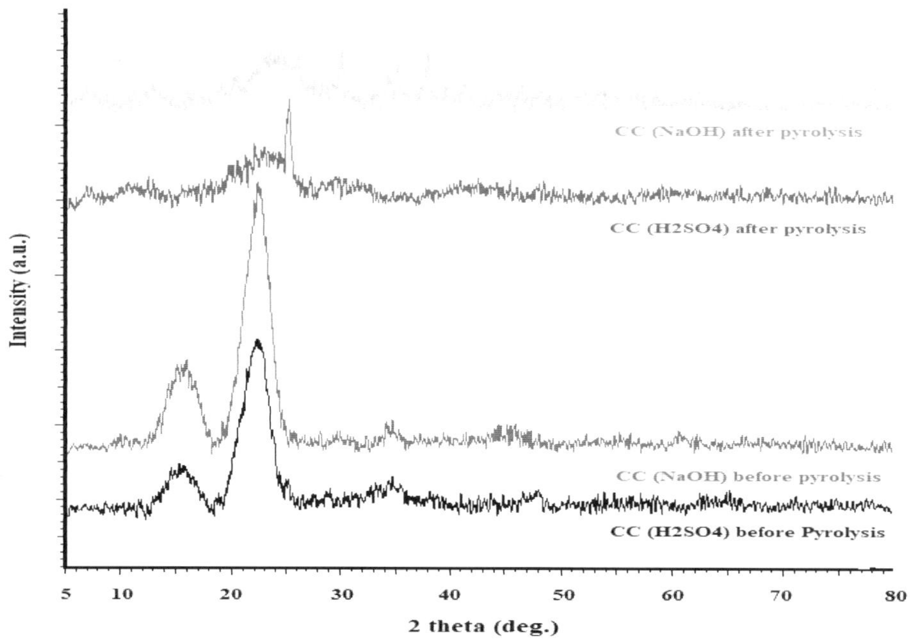

Figure3. XRD patternt for CC prepaed samples

4. Catalytic Evaluation

The photocatalytic activity of the prepared catalyst was tested by methylene blue (MB) degradation under a set of conditions. The degradation of MB dye was observed by UV/vis spectrophotometer (Shimadzu, Model 2450). MB dye solution was used as an objective pollutant because having made is possible for the UV-Vis spectroscopy to monitor the degradation of MB dye. The calibration curve of MB dye was recorded at a wavelength 665 nm. The MB dye solution was irradiated using UV-lamp (PHILIPS-15 Watt), and was held in a closed box. The irradiation source is placed on the top at a distance of 12 cm from the glass beaker containing photocatalyst samples. The photocatalytic degradation of MB for all photocatalyst samples was done using an initial concentration of 0.05 mM for MB suspension. Prior to illumination, 20 mg photocatalyst was added to the MB aqueous solution (20ml). The photocatalyst mixture was stirred in a dark place for 30 minutes in order to reach MB absorption-desorption equilibrium needed for the commencement of the photocatalytic reaction. After that, the photocatalyst mixture was exposed to a UV-lamp for a maximum irradiation time of 120 minutes. The photocatalyst mixture was withdrawn and filtered in order to separate the photocatalyst powder from the MB dye solution. The MB degradation was calculated according to the Beer-lamber Law [6]. The concentration of MB is proportional to absorbance of MB. Hence, the degradation efficiency of MB can be calculated by following equation (1):

$$\text{Degradation } (\%) = (C_o - C_t)/C_o \times 100 \qquad (1)$$

Where, C_o is the initial concentration of MB solution and C_t is concentration of MB after 120 minutes' irradiation time. The percent (%) of MB degradation calculated by the above equation is an important value that can be used to measure the catalyst performance for each sample (Table 1).

Table 1. The photocatalytic performance result for prepared samples in different weight ratio

sample	**5% TiO2**	**10% TiO2**	**5% CC(H2SO4) Before pyrolysis**	**10% CC(H2SO4) Before pyrolysis**	**5% CC(H2SO4) After pyrolysis**	**10% CC(H2SO4) After pyrolysis**	**5% CC(NaOH) After pyrolysis**	**10 % CC(NaOH) After pyrolysis**
Degradation	97.52	97.36	99.73	98.2	94.94	94.2	92.78	97.16

The result revealed although the photocatalytic performance of impregnated TiO_2 on acid treated CC is improved compared to bare TiO_2, the pyrolysed acid and base treated samples exhibit lower performance. According to XRD results, decreasing of catalytic performance of pyrolysed samples could likely be attributed to the interaction of nanoTiO_2 and treated CC during pyrolysis which leads to lower crystallinity of TiO_2. Intercalated nano TiO_2 with lower crystallinity decrease the active sites for photocatalytic reaction and decrease catalytic performance.

5. Conclusion:

Although the initial investigation of utilizing the modified coconut coir as green substrate showed better performance compared to bare nano TiO_2, further investigation is needed to reveal the reason behind lower performance of pyrolysed samples which showed higher porosity and higher thermal stability. Applying impregnation procedure after pyrolysis of treated CC could be better option for increasing the photocatalytic performance of samples. Currently, further studies for improvement of CC as green substrate in catalytic application is in progress.

References:

[1] Gawande, P.R., *Preparation and activation of activated carbon from waste materials-A review.* IJRASET International Journal for Research in Applied Science and Engineering Technology, 2016. **4**(XII): p. 1-4.

[2] Yahya, M.A., Z. Al-Qodah, and C.W.Z. Ngah, *Agricultural bio-waste materials as potential sustainable precursors used for activated carbon production: A review.* Renewable and Sustainable Energy Reviews, 2015. **46**(Supplement C): p. 218-235.

[3] Sreńscek-Nazzal, J., et al., *Production, characterization and methane storage potential of KOH-activated carbon from sugarcane molasses.* Industrial Crops and Products, 2013. **47**: p. 153-159.

[4] Emdadi, Z., et al., *Chemically Treated Rice Husk Blends as Green Desiccant Materials for Industrial Application.* Chemical Engineering & Technology: p. n/a-n/a.

[5] Suhas, P.J.M. Carrott, and M.M.L. Ribeiro Carrott, *Lignin – from natural adsorbent to activated carbon: A review.* Bioresource Technology, 2007. **98**(12): p. 2301-2312.

[6] Afifah, N., et al., *Photodegradation of methylene blue by LaFeO₃/ZnO nanocomposites under visible and UV light irradiation.* J. Phys. Conf. Ser. Journal of Physics: Conference Series, 2017. **820**(1).

IOP Conf. Series: Materials Science and Engineering **293** (2017) 012011 doi:10.1088/1757-899X/293/1/012011

Modelling the Solid Waste Flow into Sungai Ikan Landfill Sites by Material Flow Analysis Method

Latifah A. Ghani[a*], Nora'aini Ali[b] and Nur Syafiqah A.Hassan[c]
[a]School of Social Development and Economic, Universiti Malaysia Terengganu, 21030 Kuala Nerus, Malaysia,

[b,c] School of Ocean Engineering, University Malaysia Terengganu, 21030 Kuala Nerus, Terengganu, Malaysia

[a*]latifah.ghani@umt.edu.my

Abstract. The purpose of this paper is to model the material flow of solid waste flows at Kuala Terengganu by using Material Flow Analysis (MFA) method, generated by STAN Software Analysis. Sungai Ikan Landfill has been operated for about 10 years. Average, Sungai Ikan Landfill receive an amount around 260 tons per day of solid waste. As for the variety source of the solid waste coming from, leachates that accumulated has been tested and measured. Highest reading of pH of the leachate is 8.29 which is still in the standard level before discharging the leachate to open water which pH in between 8.0-9.0. The percentages of the solid waste has been calculated and seven different types of solid waste has been segregated. That is, plastics, organic waste, paper, polystyrene, wood, fabric and can. The estimation of the solid waste that will be end as a residue are around 244 tons per day.

1. Introduction
The inclining of the population in Malaysia did not just increasing the level and the rate of solid waste produced per day of per capita. Other than incineration, composting, burning, solid waste typically will be sent to landfill. Based on CIA Factbook on Malaysia, as estimated in July 2016, the population of Malaysia are reaching 30,949,962 people across the country. A studies that were run by government on last year showed that the average Malaysian produced almost 1 kilogram of solid waste per capita per day [1]. Referring to the number of Malaysian's population on average solid waste produced per person, it is not impossible to predict that the amount of solid waste produced will rise up to 30, 000 tons per day by year 2020 [2]. Therefore, the main objective of this study focused on to analyzed amount and composition of solid waste at the Sungai Ikan Landfill that will end up as residues. Other than that, this study will also provide us with the percentage of solid waste portion and characterization of leachate composition currently.

2. Material and Method

2.1. Area of Study
Sungai Ikan Landfill (SIL) has been operated for almost 10 years and now categorized as level 4 landfill (Sanitary landfill with leachate treatment facilities). It has been upgraded and transformed by Hexagon Energy. At the early operation of SIL, crude dumping and burning became the cause of

IOP Conf. Series: Materials Science and Engineering **293** (2017) 012011 doi:10.1088/1757-899X/293/1/012011

concern among the residents because of the close parameter with Sungai Nerus, Terengganu, Malaysia. What makes the condition become worst is during the monsoon season, leachate that came from the landfill, contaminated the river. In the year 2014, in history of Malaysia, one of the worst flood has occurred, which has risen the concern of SIL where efforts were made to make an eco-friendly landfill by using geotextile to trap leachate and other contaminants. The new design that were managed by Hexagon Synergy, includes the constructions of leachate collection system, a recycling central and tidal gate [3].

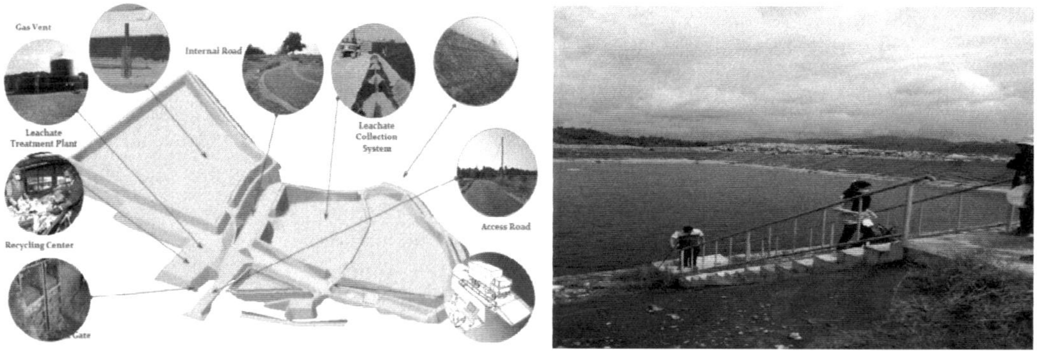

Figure 1. Sungai Ikan Landfill illustrated by Hexagon Energy with the Leachate Retention Ponds

2.2. MFA Methods
Material flow analysis comprises several elements as been listed out [4]: a system analysis comprising goods and processes, determination of the mass fluxes of all the goods per unit of time, determination of the concentration of the selected elements in this goods, calculation of the mass and fluxes from the mass fluxes of goods and element concentrations in these goods and finally, interpretation and presentation of the result.

2.3. STAN Software Analysis
After building a graphical model, with a predefined components in the software itself (process, flows, systems boundary, text field), users can enter or import known data or precise data (stocks, concentrations, mass flows, transfer coefficients) for different layer (substance, good, energy) and period to calculate an unknown quantities. All flows can be displayed in Sankey-style, in example, the width of the flow is proportional to its value. The graphical picture of the model can be printed or exported. For data export and import, Microsoft Excel is used as an interface. To build the model MFA-STAN, components consist of a system boundary, processes and flows [5].

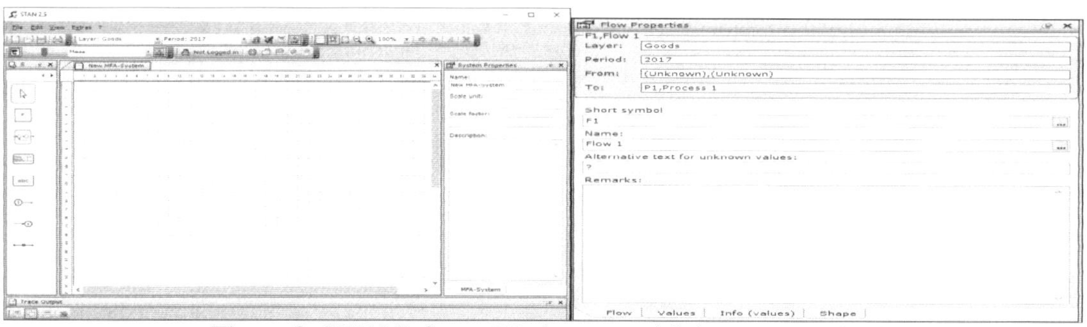

Figure 2. STAN Software Workspace and Subsystem Involved.

2.4 Experimental Analysis of Leachate Water
Sample of leachate were collected and several analysis readings have been recorded. There are two type of analysis that could be used. First, 'in-situ' analysis, and second is 'ex-situ' analysis.

3. Result and Discussion

3.1 Solid waste composition

Both result and discussion from the result were analyze. Segregation percentage, leachate characterization and MFA using STAN software were calculated and displayed. As for the solid waste that were sent to this landfill, a rough calculation has been made by using acquired data from the person in-charge at the Sungai Ikan Landfill. As the objective of this study is to determine Material Flow Analysis of the solid waste at Sungai Ikan Landfill, a segregation process is a must to determine the quantity and quality of waste composition that were deposited at the landfill sites (Table 1) [6].

Table 1. Average Weight of Solid Waste in Sungai Ikan Landfill

Type of waste	**Average weight (504.68 gram)**
	Weight (gram)
Plastics	229.65
Organic waste	191.91
Paper	102.54
Polysterene	12.72
Wood	29.8
Fabric	18.59
Can	20.58

For the record, Majlis Bandaraya Kuala Terengganu (MBKT) recorded almost 300 tons of solid were brought to Sungai Ikan Landfill a day. Which makes, SIL receive almost 9000 tons of solid waste per month. As the majority of the solid waste were brought from residential area, shop lots and larger premises, it can be concluded that most fraction of the solid waste consists of plastics and organic waste. As been projected by one of the contractor from the landfill, with this attitude, the lifespan of the landfill could be reduced to just 3 years instead of 5 to 6 years.

3.2 Material Flow Analysis

Many possible applications exist for material flow analysis (MFA) and of the function is to help with sustainable materials management (SMM) [7]. In towards sustainable solid waste management, the data from the sampling were analyzed and were inserted to the MFA calculating software called STAN, that often used by researcher. As we can observe from the Sankey diagram below, the total assumption of input that have been carried to Sungai Ikan Landfill can reach up to 88,000 tons a year. After the process of sorting at landfill, certain of recyclable item were separated and non-recyclable item such as organic waste, wood, polystyrene and fabric were left to be residue at the Sungai Ikan Landfill as depicted in MFA model in Figure 3.

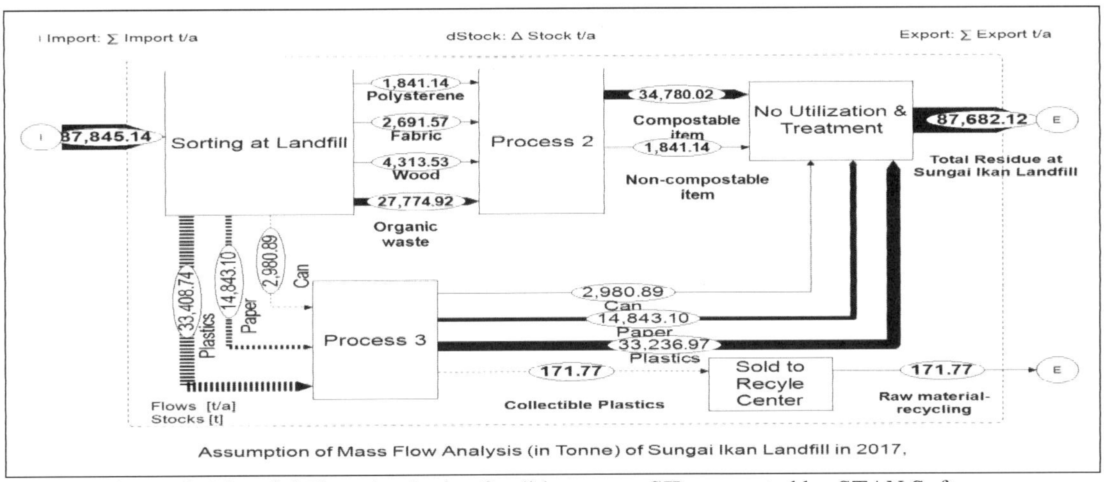

Figure 3. Material Flow Analysis of solid waste at SIL generated by STAN Software

3.3 Leachate characterization

The table 2, below are the reading of the parameters, taken for 3 days.

Table 2. Leachate's Characterization at SIL (In-Situ Analysis)

Parameters	Day 1	Day 2	Day 3
Conductivity	1.92	2.02	1.98
DO (mg/L)	1.36	1.23	1.31
Temperature (℃)	29.0	30.9	29.4
TDS	1160.3	1174.0	1163.5
PH	8.29	8.08	8.19

4. Conclusion

From the result and discussion shown above, it can be concluded that Material Flow Analysis systems can be one of the effective method in conducting environmental assessment of the solid waste residues at the landfill. Considering the large amount of residues can be calculated using the projection using STAN Software analysis. People of Malaysian are still not aware of the consequences from not separating their waste. Other than that, the fraction and percentages of solid waste that were brought to SIL were able to be identified. Total residues at SIL can reach up to 88,000tones per year. With this amount of solid waste, Malaysian people should be aware that one day, we will run out of space to be converted to landfill. The sustainable waste planning and management is very important for this consequences.

References

[1] Izni Izzati M, Mohd Zulkifli Mohd Y and Fatimah SA, *Malaysian Journal of Civil Engineering 28 Special Issue* 2016 **3** p. 180-191.

[2]Rajkumar N, Subramani T, and Elango L,. *Journal of Environmental Health Science & Engineering* 2012 **9**, 35.

[3] Agamuthu and Fauziah. *Journal of Material Cycles and Waste Management* 2010 **3** 1-2.

[4] Spyridon Achinas. *Journal of Sustainable Development Studies* 2014 **5**, 29-39.

[5] Annisa B. *In Proceedings ACES :Annual Civil Engineering Seminar.* 2016 **1**, 235-242.

[6] Agamuthu P, Venu Mahendra MS, and Mohd Afzanizam M. *Malaysian Journal of Science* 2011 30 **1**, 16-27.

[7] Oliver C, and Helmut R., 2008. *Material Flow Analysis with Software STAN.* Environmental Informatics and Industrial Ecology p. 21-80.

IOP Conf. Series: Materials Science and Engineering **293** (2017) 012012 doi:10.1088/1757-899X/293/1/012012

Homogenization Effect on Nanostructure and Conductivity of Polyaniline Nanofibre Synthesis by Mini-Emulsion Polymerization Technique

M Mohammad[1], S Kamarudin[1], N H Mohamed[2], N Asim[1], K Sopian[1]

[1]Solar Energy Research Institute (SERI), Universiti Kebangsaan Malaysia, 43600, Bangi, Selangor, Malaysia
[2]Malaysia Nuclear Agency (Nuclear Malaysia), 43600, Kajang, Bangi, Selangor, Malaysia

masita@ukm.edu.my

Abstract. Nanofibre polyaniline (n-PANI) was synthesized by mini-emulsion polymerization technique between aniline monomer and ammonium persulfate as an oxidant using homogenizer. The synthesis was performed by optimizing mixing speed from 10,000 to 30,000 rpm and time reaction between 0.5 to 24 hours at fixed monomer to oxidant molar ratio 4:1. An attempt has been made to investigate on how the speed of homogenizer affects the size and conductivity of n-PANI. The formation of n-PANI chain was confirmed by Fourier transform infrared spectroscopy (FTIR). The X-ray diffraction (XRD) spectra revealed PANI crystalline nature. Hall effect measurement used indicated that the electrical conductivity of n-PANI is increased with homogenizer speed from 5.2 to 17.5 Scm-1. The morphological properties of n-PANI performed by scanning electron microscopy (SEM) show the decreasing size of n-PANI from 50-60 nm to 20-30 nm with the increment homogenizer speed. This study indicated the optimum speed parameter of homogenizer play a role in reducing the nanostructured size and thus, increasing the electrical conductivity of n-PANI.

1. Introduction

Polyaniline is one of the most intensively studied conducting polymers with a wide potential application as a functional polymer [1, 2]. Nanofibre polyaniline (n-PANI) is one of the promising conducting polymers due to its environmental stability, controllable electrical properties and high surface area [1]. Electrochemical and chemical methods are the common techniques to synthesized polyaniline and its derivative. Recently, the interests in nano structured PANI have increased tremendously. The mini-emulsion polymerization is known to be a prevailing technique to produce a whole variety of conducting polymer. As compared to the conventional method such as electroplating, mini-emulsion polymerization technique capable of controlling over reaction rates and heat dissipation by applying high shear stress using ultrasonication or high-pressure homogenizer technique [3, 4, 5].
Crystallinity is an indication of the amount of crystalline region in the polymer. High crystallinity will result in high conductivity. The polymer obtains by mini-emulsion method is high crystallinity compare to conventional method [3]. Polyaniline exists in various oxidation states such as pernigraniline, leucoemeraldine and emeraldine. Leucoemeraldine which is fully reduced is environmentally unstable and very reactive even with small amount of oxygen. While pernigraniline is fully oxidized which caused fully imine group and tend to not oxidized, hence environmental friendy.

Emeraldine which is partially oxidized is chemically and environmentally stable. However, the only emeraldine is conducting while pernigraniline and leucoemeraldine are both an insulator. [6].
This surfactant-free emulsion polymerization method is a facile technique to prepare the PANI nanofibers for higher yield and conductivity without application of surfactant. This technique also provides uniform and well-distributed n-PANI without contaminating of surfactants. The surfactant is a substance that lower the surface tension between two liquid. Surfactants such as dodecyl benzene sulfonate (DBSA) and Sodium Dodecyl Sulphate (SDS) are the common surfactant in the polymerization of polyaniline [7]. However, the post-process of removing the surfactant will be an issue in order to get high purity of PANI.
This research aims to study the electrical conductivity and size of n-PANI nanofibers at different homogenizing speed and time reaction. The FESEM, XRD and FTIR techniques were used to analyse the properties of n-PANI.

2. Materials
Aniline (R&M), Chloroform (R&M), nitric acid 70% (QREC), Ammonium persulfate (R&M) and Acetone (QREC). All chemicals are in analytical grade.

3. Methodology
The n-PANI was prepared by mini-emulsion polymerization technique. Typically, in 500 ml conical flask, 9.3g aniline was dissolved in 200 ml chloroform (CHCl3, analytical grade). The flask was covered with parafilm. In another beaker, 5.7g of ammonium persulfate (APS) was dissolved in 200 ml, 1M nitric acid and the beaker was covered with parafilm to avoid evaporation. A mixture of aniline with chloroform and APS with nitric acid were stirred moderately with a magnetic bar for 30 minutes in a separate beaker. After 30 minutes, APS was transferred into aniline solution. These two solutions were mixed rapidly in exactly 3 minutes by homogenizer with the different speed which was 10 000, 20 000 and 30 000 rpm respectively to study the effect of homogenizer speed in nanofiber formation. Mixing with magnetic stir was done as a reference. After homogenized, the flask was covered with parafilm with no stirring to get the interfacial polymerization effect and kept at room temperature in a comparison time which was 0.5 1, 2, 4, 6, 15 and 24 hours. After specific hours, 100 ml acetone was added to the flask to stop the polymerization. The product was filtered and washed carefully with 0.1M nitric acid and followed with acetone until the filtrate becomes colourless and finally dried in oven 40° C for 24 hours [8-10].
The n-PANI will be characterized by Fourier transform infra-red (FTIR) Spectroscopy, (Spectrum 2000 Explorer; Perkin Elmer Cetus Instruments, Norwalk, CT) to confirmed and analyse the functional groups present in the n-PANI. The n-PANI pellet for x-ray diffraction measurements was prepared by using manual hydraulic pressure using 15-tonne pressure. Powder X-ray diffraction pattern was recorded by DS Advanced, Bruker. TGA Thermal gravimetric analysis of the n-PANI prepared under the optimal condition was carried out using thermogravimetry analyser Perkin Elmer Pyris1 to study n-PANI thermal stability. About 4 mg of n-PANI was weighed into an alumina crucible and the thermogram was recorded from 25 °C to 800 °C at a heating rate 10 °C/min under 20 mL/min N2 flow. The morphology of the n-PANI was examined by FESEM (GEMINI, ZEISS). Room temperature electrical conductivity measurements were carried out by and hall effect study on compressed n-PANI pellets. Electrical conductivity was measured by Ecopia Hall Effect Measurement System (Ecopia, HMS).

4. Result and discussion

4.1. FTIR
n-PANI was successfully prepared by mini-emulsion polymerization technique. The FTIR spectra of n-PANI prepared under the speed of 30000 rpm via homogenizer with 2 hours time reaction is shown in Figure 1. The main peaks consistent with C=N and C=C stretching vibrations of quinone and benzene ring [8] could be observed at around 1574 cm−1 and 1489cm−1 which confirmed the

presence of n-PANI [9, 10]. The peaks at 1372 cm−1 and 1291 cm−1 are attributed to the C-N stretching mode and the peak at 1114 cm−1 reveals the C-H bending vibrations [10].

4.2. XRD

Commonly, a polymer has an amorphous structure, however, n-PANI is showing crystalline structure due to their fibre nature and planar nature of benzenoid and quinoid functional groups [10]. Figure 2 shows the X-ray diffraction pattern of n-PANI emeraldine salt prepared under the speed of 30000 rpm via homogenizer with 2 hours time reaction forming three peaks. The characteristic peaks appeared at $2\theta = 14.8°$, $20.5°$ and $25°$, corresponding to (011), (020) and (200) crystal planes of PANI [8, 12]. These peaks are ascribed to the periodicity parallel and perpendicular to the polymer chains of PANI, respectively [13].

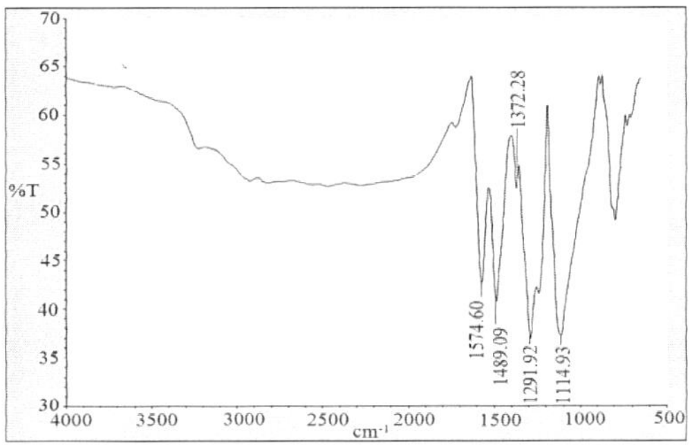

Figure 1. FT-IR spectrum of n-PANI

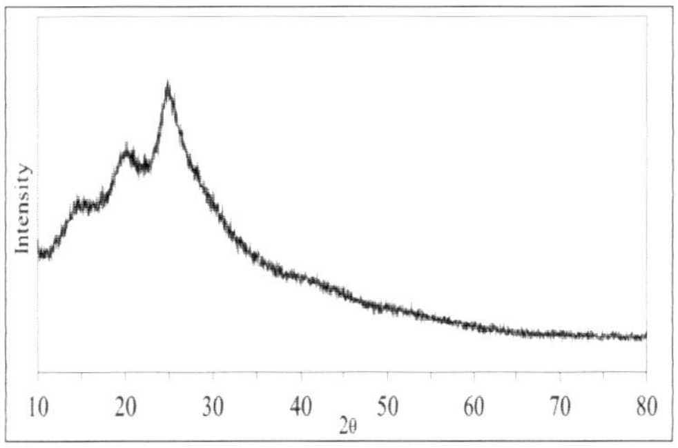

Figure 2. XRD diffractogram of n-PANI

4.3. Thermogravimetry analysis

Thermogravimetric analysis (TGA) of the n-PANI synthesized by mini-emulsion showed good stability at the higher temperature. Figure 3 shows thermogram of n-Pani prepared under the speed of 30000 rpm via homogenizer with 2 hours time reaction. The little weight loss in the initial stages around 6% is due to the volatilization of water and small molecules (oligomers and unreacted monomers) [14].

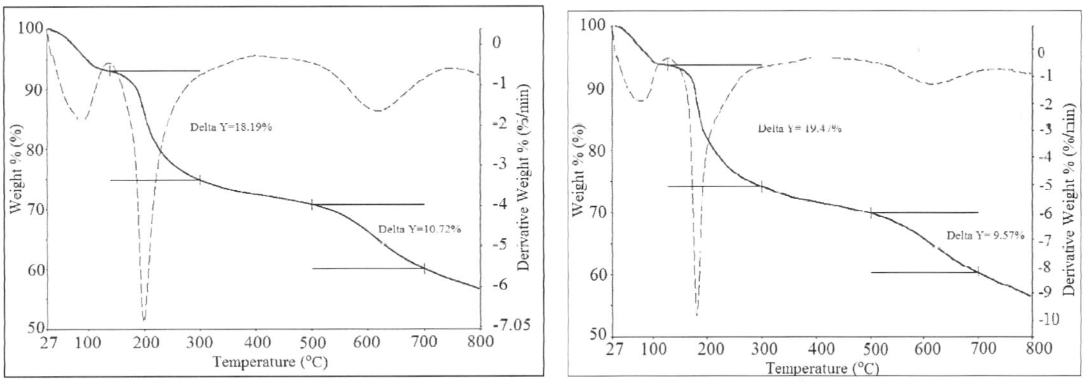

a b

Figure 3. TGA thermograph of n-PANI a) without homogenizer b) with homogenizer

However, severe weight loss with a large slope occurring among 140-280°C for PANI mostly originates from the dopant acid [15, 16]. The weight loss among 250-550 °C is caused by the degradation of the n-PANI main chain. The total weight loss at 800 °C was around 60%. Both n-Pani prepared by the magnetic stirrer and homogenizer shows similar patent. However, for n-PANI prepared by the homogenizer, it gives extra weight loss at 140-280°C may due to the higher amount of dopant acid it have compared to n-PANI prepared by magnetic stirring.

IOP Conf. Series: Materials Science and Engineering **293** (2017) 012012 doi:10.1088/1757-899X/293/1/012012

4.4. Morphology

The homogenizing effect on the morphology of n-PANI synthesized was investigated by FE-SEM. As shown in Figure 4, the nanofiber size between the polymeric aggregates is decreasing as synthesis temperature decreasing. In this way, an improvement of size due to the reducing of droplet size by homogenizer lead to a reduction of nanofiber size. Similar to TGA result, increasing of homogenizer speed may increase surface area by reducing of droplet size and results in an increment of the amount of dopant which related to the conductivity improvement.

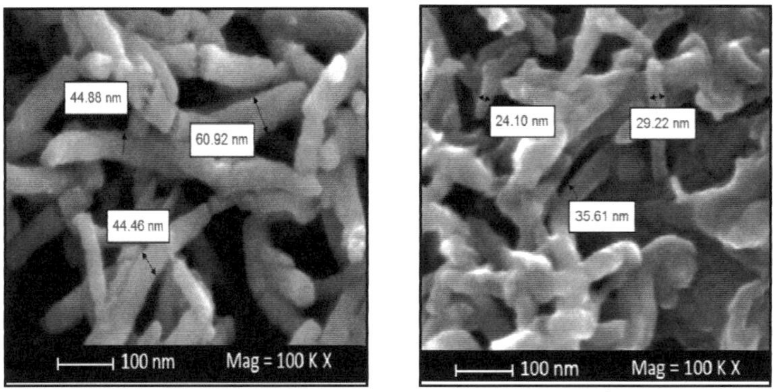

a b

Figure 4. FESEM photograph of n-PANI a) without homogenizer b) with homogenizer

4.5. Electrical Conductivity

The electrical conductivity was measured by Hall effect technique. The influence of homogenizing effect on the conductivity as shown in Table 1 shows increasing of conductivity by increasing the homogenizer speed. The formation of nanofibers takes place in nucleation and initial growth stages. By applying high shear stress from homogenizer, the size of the droplet in nucleation step will decrease. Then the initial stage continues with smaller nucleation site which results in the formation of smaller nanosize compared to n-Pani without homogenizing effect. With the increasing formation of n-PANI nanofiber, which increased the dopant site resulting in increasing of electrical conductivity. Effect of time reaction on the electrical conductivity of n-PANI prepared using 30000 rpm of homogenizer speed is shown in Table 2. By the time of reaction, oxidation of emeraldine will reduce to completely polymerize. After 2 hours reaction, the electrical conductivity of n-PANI decreases and become stable up to 24 hours. It may explain by completely oxidized polyaniline that would be in the pernigraniline state. While reaction time less than 2 hours, aniline may not completely turn to emeraldine state which results to lower electrical conductivity.

Table 1. Effect of stirring speed

Speed(rpm)	Conductivity (Scm-1)	Size
Magnetic	5.2	55-65
10000	12.3	50-55
20000	15.9	35-40
30000	17.5	25-30

Table 2 Effect of time reaction

Time (Hour)	Conductivity (Scm-1)	Size
0.5	6.3x 10-3	25-30
1	1.5x10-2	25-30
2	17.5	25-30
4	15.6	30-35
6	10.4	35-45
15	11.1	35-45
24	11.2	35-45

5. Conclusion

The (n-PANI) was successfully synthesized by mini-emulsion polymerization technique. The optimum speed and time reaction of the homogenizer for polyaniline polymerization to obtain the highest conductivity with reducing size of nanofiber is 30000 rpm in 2 hours. Electrical conductivity was strongly influenced by time reaction which longer period promotes to oxidation hence reduce the conductivity.

Acknowledgments

The authors greatly acknowledged the financial support by FRGS/1/2014/TK04/UKM/03/2 and GGPM-2014-027 from CRIM (Centre of Research and Instrument Management, UKM and KPT (Kementerian Pengajian Tinggi), Malaysia.

References

[1] Hong Y, Hwang S, Yoon D S. and Yang J 2015 *Sensor Actua.t B-Chem.* **218**, 31–36.
[2] Shamagsumova R, Porfireva A, Stepanova V, Osin Y, Evtugyn G and Hianik T 2015 *Sensor Actua.t B-Chem.* **220**, 573–582.
[3] Marie E, Rothe R, Antonietti M and Landfester, K 2003 *Macromolecules*, **36**(11), 3967–3973.
[4] Wang H, Wang L, Wang R and Tian X 2011 *J. Mat. Sci.* **46**(4), 1049–1052.
[5] Elhalawany N, Elmelegy H and Nayfeh M 2015 *Synth. Met.* **205**, 145–152.
[6] Stejskal J and Gilbert R G 2006 *Pure Appl Chem.* **74**(5) 857-867
[7] Kumar A K, Mishra A, Awasthi K and Kumar V 2015 *Macromol. Symp.* **357**, 168–172
[8] Abdolahi A, Hamzah E, Ibrahim Z and Hashim S 2012 *Materials*, **5**, 1487–1494.
[9] Shao L, Qiu J, Liu M, Feng H, Lei L, Zhang G and Zhao Y 2011 *Synthetic Met*, **161**(9-10), 806–811.
[10] Kavitha B, Prabakar K, Siva K, Srinivasu D, Srinivas C, Aswal V K and Siriguri V 2012 *IOSR J Appl Chem*, **2**(1), 16–19.
[11] Vivekanandan J, Ponnusamy V, Mahudeswaran A and Vijayanand P S 2011 *Appl Sci Res*, **3**(6), 147–153.
[12] Bhadra S and Khastgir D 2008 *Polym Test,* **27**(7), 851–857.
[13] Srinivas C H, Srinivasu D, Kavitha B, Narsimlu N and Kumar K S 2012 *IOSR J Appl Phys* **1**(5), 12–15.
[14] Mahmood W A K, Rahman K M M, Yeow Y G and Wee Y K 2012 *Int J. Chem Reac Eng,* **10**(1). 1-20
[15] Masdarolomoor F 2006. *Novel nanostructured conducting polymer systems based on sulfonated polyaniline*. PhD Thesis. University of Wollongong. Australia
[16] Wang Y, Chen K, Li T, Li H, Zeng R, Zhang R, Gu Y, Ding J and Liu H 2014 *Synthetic Met,* **198**, 293–299.

IOP Conf. Series: Materials Science and Engineering **293** (2017) 012013 doi:10.1088/1757-899X/293/1/012013

Adsorption Isotherm of Chromium (Vi) into Zncl$_2$ Impregnated Activated Carbon Derived by Jatropha Curcas Seed Hull

M Mohammad[1], I Yakub[2], Z Yaakob[3], N Asim[1], K Sopian[1]

[1]Solar Energy Research Institute (SERI), Universiti Kebangsaan Malaysia, 43600, Bangi, Selangor, Malaysia
[2]Department of Civil Engineering, Fakulti Engineering, Universiti Malaysia Sarawak
[3]Department of Chemical and Process Engineering, Faculty of Engineering and Built Environment, Universiti Kebangsaan Malaysia

masita@ukm.edu.my

Abstract. Hexavalent chromium is carcinogenic and should be removed from industrial wastewater before discharged into water resources. Adsorption by using activated carbon from biomass is an economic and conventional way on removing the heavy metal ions from wastewater. In this research, activated carbon is synthesized from *Jatropha curcas* L. seed hull through chemical activation with $ZnCl_2$ and carbonized at 800 ^oC (JAC/$ZnCl_2$). The activated carbon has been characterized using FTIR, SEM-EDX, BET and CHNS-O analyzer. Adsorption isotherms have been analysed using Langmuir and Freundlich models to determine its removal mechanism. The maximum adsorption capacity of Cr (VI) metal ions onto JAC/$ZnCl_2$ activated carbon is 25.189 mg/g and following Langmuir isotherm model which is monolayer adsorption.

1. Introduction

Chromium exists in almost anywhere on earth, including seawater and plants, but its hexavalent state (+6) has been proven to be the most poisonous and carcinogenic to both human and animal [1]. Because of its high corrosion-resistant and hardness, chromium has been used in many metal industries, ranging from printing machine to aerospace manufacturing. Many researchers have tried other materials to replace this element in the industry but none of the materials can surpass the excellent layer quality and cost-effectiveness it holds [2].

Hexavalent chromium can be reduced to trivalent chromium that is insoluble in water and can be separated easily [3]. This reaction is manipulated by many of the removal techniques in industrial wastewater, including acidification with sulphuric acid, chemical reduction with sulphite sulphur compounds and neutralization with a sodium hydroxide solution. The other way is through a separation process such as ion exchange and membrane [4]. However, these methods produce a huge amount of sludge that will end up in the landfill, which creates another environmental concern [2]. Besides ion exchange and membrane separation techniques, adsorption method has attracted many attentions nowadays [5]. This is because of the ability to recover and reuse precious chromium that is adsorbed using activated carbon, instead of contributing to soil and river pollution by sending the waste sludge to landfills. Some of the materials that have been used as an adsorbent of chromium (VI) were marine algae, KNb3O8 compound and sol-gel hydrotalcite-like compounds [5, 6, 7]. Besides,

Shah et al. (2009) have combined both methods of reduction of Cr (VI) to Cr (III) with sorption using natural sorbent material to increase the efficiency of chromium removal [8].
Activated carbon is synthesized from carbonaceous materials, called the precursor that mutually affecting the adsorption performance of the produced activated carbon with its activation properties [9]. There are two types of activation method namely physical and chemical activations. Zinc chloride ($ZnCl_2$), being one of the most common chemicals used in chemical activation, has several reported advantages towards physical activation. Ahmadpour and Do [10] found that it can improve the pore creation and gives higher yield because the chemical retards tar escape during carbonization and endorses the condensation reaction. Mohanty et al. (2005) had also observed that $ZnCl_2$-activation increases BET surface area [10].
Since the material cost of activated carbon is high, activated carbons produced from a low-cost material such as biomass have brought special attention to researchers in order to develop a sustainable wastewater treatment in the industry [11, 12]. Tumeric waste, Ferronia shell, Delonix shell and Ipomea Carnia stem are among other biomass that have been studied and proven to have potentials of being activated carbon [13]. Esmaeili et al. (2010) have used algae as the precursor for the activated carbon to remove chromium from solution while Tangjuank et al. (2009) have used cashew nut shells [5,14]. Besides, Venkatraman et al. (2009) have produced low-cost acid-activated activated carbon from Borassus bark to adsorb chromium ions from metal solution [15].
Seed hull of *Jatropha curcas L.* or physic nut is one of many biomasses that can be used as a precursor. It is mainly produced from soap and medicine industry as well as the bioenergy production [16]. Many studies have been done to use a physic nut as a precursor of activated carbon, one of which is done by Namasivayam et al. (2007) where they used the husk to adsorb anions, heavy metals, organics and dyes from water [17]. Mohammad et al. (2010) studied the removal of cadmium and zinc ions from aqueous solution using a physic seed hull and concluded that it can be a potential low-cost adsorbent for the removal of metal ions [18]. This project synthesized activated carbon from jatropha seed hull by chemical activation technique with Zinc chloride (JAC/$ZnCl_2$). The JAC/$ZnCl_2$ has been utilized as an adsorbent for chromium ion removals from aqueous solutions. The experimental data were fitted to the Langmuir and Freundlich isotherm models to determine its isotherm mechanism.
This research aim is to improve the surface area and porosity of jatropha seed hull by chemical activation, hence increase its adsorption capacity for chromium metal ions in aqueous solution. The modification was improved its surface area and reduce the pore size, consequently, improved it adsorption tendency and capacity as compared with research by Mohammad et al. (2010) [18].

2. Materials and Methods

2.1. Synthesis of Jatropha Activated Carbon

All chemical used were analytical grade and were obtained from Merck. Jatropha seed hull (JSH) was washed and soaked in 80°C water until clean. Then, it was dried in 80°C oven for 24 hours and later soaked in $ZnCl_2$ overnight. The $ZnCl_2$-activated JSH was dried before carbonized in MTI Corp OTF-1200X (USA) tube furnace at 10°C/min for 60 minutes at 800°C under Nitrogen flow. After cooled, they were drenched in 1 M HCl solution at 80°C and rinsed with distilled water until the pH becomes 6. Finally, JAC/$ZnCl_2$ is sieved and -0.5+0.25 mm sieve size of the particle is used as an adsorbent.

2.2. Characterization of Jatropha Activated Carbon

The characteristics of JAC/$ZnCl_2$ were observed and analyzed through its elemental composition, inorganic components, surface morphology, surface area, pore volume, porosity and surface functional groups. The elemental composition was analyzed using Thermo Fin 1112 CHNSO-Analyzer (Italy) and the inorganic components were examined in PANalytical Mini Pal 2 X-Ray Fluorescence (Netherlands). The surface morphology was studied using CARL ZEISS EVO MA10 Scanning Electron Microscope (UK). The surface area and the porosity characteristic were investigated using BJH method with BET Micromeritics ASAP 2020 (USA). The surface functional groups were detected using Thermo Scientific Nicolet 6700 Fourier Transform Infrared Spectroscopy (USA).

2.3. *Adsorption Isotherm*

The adsorption of chromium by $JACZnCl_2$ was carried out in a batch system. Chromium solution of 1,000 ppm was prepared by using 2.83 g of potassium dichromate, K2Cr2O7 in 1 L distilled water. Then, aqueous solutions of 25 mL with the initial concentration in the range of 5 to 60 ppm were prepared from the 1,000 ppm solution. All batches were done with replicates in 100 mL beaker and capped with aluminium foil. They were also rotated on LABWIT ZHWY-304 (China) rotary shaker at 160 rpm after 0.1 g of adsorbent was added. Each sample taken is diluted to an appropriate ratio before being analyzed using Perkin Elmer Analyst 800 Atomic Absorption Spectrometer (USA).

2.4. *Mathematical Models*

Adsorption capacity, q_t

$$q_t = \frac{(C_0 - C_t)V}{m} \tag{1}$$

where, q_t is the amount of chromium adsorbed per gram of $JAC/ZnCl_2$ (mg g^{-1}), C_0 and C_t are the initial and final chromium concentration in the solution for time *t* (mg L^{-1}).

Adsorption capacity at equilibrium, q_e

$$q_e = \frac{q_{max} K_L C_e}{1 + K_L C_e} \tag{2}$$

$$q_e = K_F C_e^{1/n} \tag{3}$$

Where, q_e is the amount of chromium adsorbed per gram of $JAC/ZnCl_2$ at equilibrium (mg g^{-1}), C_e is the chromium concentration in the solution at equilibrium (mg L^{-1}), K_L is Langmuir constant that corresponds to adsorption energy (L mg^{-1}), q_{max} is maximum adsorption capacity (mg g^{-1}), K_F is Freundlich constant that indicates the relative adsorption capacity ($mg^{1-(1/n)}$ $L^{1/n}$ g^{-1}), and *n* is related to the intensity of adsorption. Equations (2) and (3) can be linearized into equations (4) and (5) respectively;

$$\frac{1}{q_e} = \left(\frac{1}{K_L q_{max}}\right)\frac{1}{C_e} + \frac{1}{q_{max}} \tag{4}$$

$$\ln q_e = \ln K_F + \frac{1}{n}(\ln C_e) \tag{5}$$

Separation factor, R_L

$$R_L = \frac{1}{1 + K_L C_0} \tag{6}$$

Where, R_L is separation factor (dimensionless), K_L is Langmuir constant (L/mg), and C_o is the concentration of chromium in the bulk fluid (mg/L).

3. Results and discussion

3.1. *Characterization of Jatropha Activated Carbon*

The elemental analysis showed that carbon dominates by 65% of the weight of $JAC/ZnCl_2$, followed by oxygen, hydrogen and nitrogen with a composition of 31%, 3% and 1% respectively. These elements are bonded to carbon either as the basic structural units in the microtexture of the activated carbon or as surface functional groups when bonded at the edge of the aromatic sheets [19]. Figure 1 shows the magnified image of $JAC/ZnCl_2$ that was used for adsorption of chromium. Before adsorption, it can be seen that it has considerably good surface morphology because of the presence of natural pores, which gives a potential of being a good adsorbent. The three groups of pores in accord with the International Union of Pure and Applied Chemistry (IUPAC) are the macro, meso and micro that have more than 50 nm, between 2 to 50 nm, and less than 2 nm in diameter respectively [20]. All of these pores can be seen distributed all over the surface with some internal pores as the zinc-activation penetrated deeper into the body of the particles. Grinding the $JAC/ZnCl_2$ to the required size has broken up the external macro and meso-pores on the surface of the particles. These breakages put together the small pores into larger ones. Hence, opened the surface to the internal meso and micro-pores, and finally increasing the surface area which enhances the capacity of adsorption. After the adsorption, chromium ions can be seen attached to the surface of $JAC/ZnCl_2$. This is validated by Energy-Dispersive X-ray scanning in Figure 1(c) where chromium ions appear on its surface.

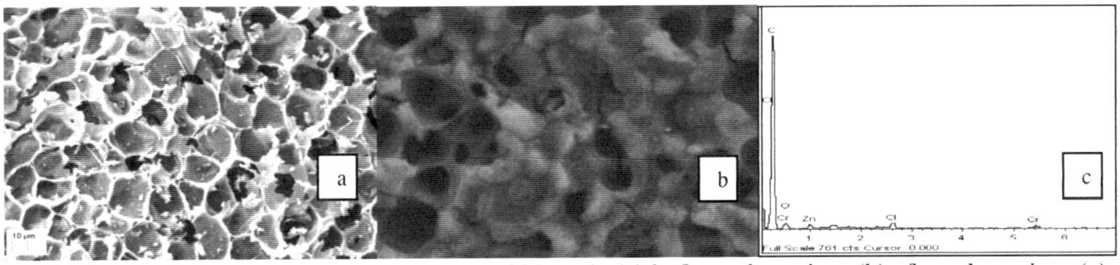

Figure 1. SEM-EDX image of JAC/$ZnCl_2$ (1000mag), (a) before adsorption; (b) after adsorption; (c) EDX analysis

Surface area is crucial for an adsorption of adsorbate onto the surface of adsorbent, and the porosity of the particles contributes to the large extent in having a larger surface area as more pores to create the bigger surface area. BET surface area for JAC/$ZnCl_2$ is 667.16 m2/g which is comparable to the Jatropha activated carbon developed by Xin-hui et al. (2010) using NaOH and Sricharoenchaikul et al. (2007) using KOH that are 1,350 and 532.3 m2/g respectively [21, 22]. Micropores contribute 69% of the total surface area of JAC/$ZnCl_2$ but the average pore diameter of 2.2 nm locates JAC/$ZnCl_2$ under mesopore class. FTIR spectra of JSH and JAC/$ZnCl_2$ has been taken and its was proved the functional groups presence before the chemical activation that is at the surface of JSH, include amines, alkynes and carboxylic acids have been completely removed in this chemical activation using $ZnCl_2$.

3.2. *Adsorption Isotherm Studies*

The adsorption isotherm can be investigated from the correlation between the adsorption capacities of chromium on JAC/$ZnCl_2$ with the concentration of chromium in the aqueous solution, both at equilibrium. The adsorption of chromium on JAC/$ZnCl_2$ for time *t* can be found from the concentration of chromium left in the solution at that time, which is shown in equation 1. Figure 2 shows the adsorption capacities of chromium for different initial concentrations against contact time until it reaches equilibrium. It shows the adsorption is increased by time and with the increment in initial metal ion concentration. It found that the equilibrium is attained at around 4 hours and followed the three-step kinetic process, starting with rapid uptake of metal ions, the uptake is slightly reduced after 20 minutes and finally, no more significant uptake which shows the pores are fully employed [18].

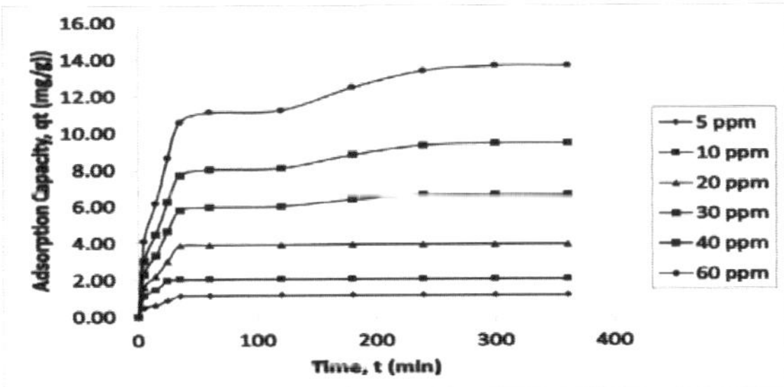

Figure 2: Adsorption Isotherm Study of Chromium with JAC/$ZnCl_2$ at 25 °C

Adsorption isotherm of chromium ions onto JAC/$ZnCl_2$ was investigated by fitting the experimental data into Langmuir and Freundlich isotherms that are described by equations 2 and 3 respectively. Equations 2 and 3 can be linearized into equations 4 and 5 respectively. Table 1 summarizes the values of Langmuir and Freundlich constants obtained from linearized equations of Langmuir and Freundlich isotherms. As Langmuir isotherm assumes that only one metal ion can be held by each site on the surface of adsorbent or monolayer adsorption, while Freundlich isotherm assumes that adsorption takes place on heterogeneous surfaces. Both R^2-values showed in Table 1 are considerably high and near unity, but Langmuir isotherm gave the best fit with the highest R^2-value. This suggests that the adsorption of chromium onto JAC/$ZnCl_2$ obeys Langmuir isotherm with maximum adsorption capacity, q_{max}= 25.189 mg/g and the affinity of the binding sites and energy of adsorption, K_L= 0.4148 L/mg. The dimensionless constant separation factor, R_L is given by equation 6. The Langmuir isotherm is favourable if the value lies between 0 and 1, while values higher than unity suggest unfavourable adsorption system. The separation factor for this system falls in the range of 0.04 to 0.33. These near-zero values depict the desired adsorption which confirms that this process is favourable [23].

Table 1: Values of Langmuir and Freundlich Constants for Adsorption of Chromium with JAC/$ZnCl_2$

Adsorbate	Langmuir constants			Freundlich constants		
	q_{max} (mg/g)	K_L (L/mg)	R^2	n	K_F (L/mg)	R^2
JAC_{ZnCl2}	25.189	0.4148	0.9906	1.3060	6.6333	0.9803

The previous works of Mohammad et al. and Sen et al., 2010, used non-activated Jatropha seed hull to adsorb metal ions from wastewater also found Langmuir isotherm satisfied the adsorption mechanism with maximum adsorption capacity of 6.724 mg/g for Zn^{2+} and 6.983mg/g for Cd^{2+} [11, 23].

4. Conclusion

The JAC/$ZnCl_2$ obtained has an improvement compared to an untreated JAC in terms of its surface area, porosity and metal ion removal from wastewater. The adsorption is monolayer on the surface of the adsorbent with maximum adsorption capacity of 25.189 mg/g, which is higher compared to untreated JAC. The adsorbent produced will be tested for its affinity towards a different type of metal ions in future. This research would benefit for the removal of heavy metal from industrial wastewater such as electroplating, solar photovoltaic and etc. in order to save the environment.

Acknowledgments

The authors greatly acknowledged the financial support by FRGS/1/2014/TK04/UKM/03/2 and GGPM-2014-027 from CRIM (Centre of Research and Instrument Management), UKM and KPT (Kementerian Pengajian Tinggi), Malaysia.

References

[1] Burrows D 1983.Chromium: *Metabolism and Toxicity*, CRC Press, USA.
[2] Frenzel I 2005. *Waste Minimization in Chromium Plating Industry*, Membrane Science & Technology. University of Twente. Enschede
[3] Watts R J 1997. *Hazardous Wastes: Sources, Pathways, Receptors*. John Wiley&Sons, USA, 1997.
[4] Cavaco S A, Fernandes S, Quina M M, Ferreira L M 2007. *J. Hazard. Mater*. 1-5.
[5] Esmaeili A, Ghasemi S, Rustaiyan A 2010. J. *Marine Sci & Technol*. **18**(4)587-592.
[6] Sinthao J, Santawitee O, Supothina S, Tawakeaw S 2010 *J. Met., Mater. & Minerals*. 133-136.
[7] Ramos-Ramirez E, Ortega N L G, Soto C A C, Gutierrez M T O 2009. *J. Hazard. Mater*. **172** (2-3) 1527-1531.
[8] Shah B A, Shah A V, Singh R R., 2009. *J. Environ. & Technol*. **6** (1) 1-13.
[9] Perrich J R 1981. *Activated Carbon for Wastewater Treatment*, CRC Press, USA
[10] Mohanty K, Jha M, Meikap B C, Biswas M N, *Chem. Eng. Sci.* (2005) **60** (11) 3049-3059
[11] Mohammad M, Sen T K, Maitra S, Dutta B K 2010. Water Air Soil Pollut. **215** (1-4) 1-12.
[12] Nunes A A, Franca A S, Oliveira L S 2009. *Bioresource Technol*. **100**(5) 1786-1792.
[13] Sivakumar S K P, Palanisamy P N 2007. *J. Chem* **5**(2) 409-426.
[14] Tangjuank S, Insuk N, Udeye V, Tontrakoon J 2009. *J. Phys. Sci.* **4**(8)412-417.
[15] Venkatraman B.R., Parthasarathy S., Kasthuri A., Pandian P., Arivoli S., 2009. *J. Chem* S1-S11.
[16] Jongschaap R E E, Corre W J 2007.Bindraban P.S, Brandenburg W.A., *Claims and Facts on Jatropha curcas L.*, Wageningen,.
[17] Namasivayam C, Sangeetha D, Gunasekaran R, 2007. Process Saf. & Environ. Protection **85** (2) 181-184.
[18] Mohammad M, Maitra S, Ahmad N, Bustam A, Sen T K, Dutta B K 2010. *J. Hazard. Mater*. **179**. 363-372.
[19] Alfarra A, Frackowiak E, Beguin F 2004. *Applied surface science*. **228**(1) 84-92.
[20] IUPAC. 1985 *Pure & Appl. Chem*. 603-619.
[21] Xin-Hui D, Srinivasakannan C, Jin-Hui P, Li-Bo Z, Zheng-Yong Z, 2010. *Fuel Process. Technol.* **92** (3) 1-7.
[22] Sricharoenchaikul V, Aht-ong C P D, Atong D, 2008, *Am. Chem. Soc.* 1-7.
[23] Sen TK, Mohammad M, Maitra S, Dutta B K 2010, *J. Clean* **38** (19) 850-858.

IOP Conf. Series: Materials Science and Engineering **293** (2017) 012014 doi:10.1088/1757-899X/293/1/012014

Research on transversal connection strengthening of T beam bridges

C Chen[1] Y Pan[2] C Q Yang[1*]

[1]School of Civil Engineering, Southeast University, Nanjing, 210096, China

[2]Jiangsu Fenguan Expressway Management Co. Ltd., Lianyungang, 222000, China

ycqjxx@seu.edu.cn

Abstract. This paper presents a new type of brace system to strengthen transversal connection of T beam bridges. A nylon T beam bridge was made to study the strengthening effect. Braces were installed in the mid span and the load was symmetrical. The results show that the reduction of the maximum deflection of the bridge with the brace system was 14.9% in experiments and 19.1% in finite element analysis, which show that the transversal connection of T beam bridge can be improved efficiently by the proposed brace system.

1. Introduction

Concrete T beam bridges have been widely used in China. They are manufactured in the precast plant and erected in the construction site. Compared with rectangle beam, T beam makes full use of compression behavior of concrete and decreases the volume of concrete in pressure region, which make it appeal for the construction of short-to medium-span bridges. With long term overload and insufficient maintenance, many damages emerged, including concrete deterioration, reinforcement corrosion and so on. One serious problem is the damage of transversal connection between adjacent beams, which causes the "single slab bearing"[1] and decreases the structural stiffness.

A number of studies have been reported on strengthening the transversal connection of T beam bridge. Xie *et al*[2] strengthened Liu Liang River Bridge by transversal prestressed tendons in Hubei, China. Field tests and finite element analysis showed that transversal prestressed tendons enhanced the transversal connection and improved the live load distribution between beams. Guo[3] strengthened the diaphragms of Guang Bei Overpass Bridge by bonding steel plates, which decreased the maximal deflection obviously and prevented crack growth. Niu *et al*[4] used the finite element method to analyze the effect of diaphragm on integrity of precast T-shape girder bridge. The results showed that the live load distribution and the integrity of the bridge were improved with the increasing number of diaphragms. Li *et al*[5] studied the effect of intermediate diaphragms on the deflection and strain of beams at mid span. It was found that the effect of diaphragms on the mid span deflection was more prominent in the short span bridge and the reduction in the maximum bending moment by the diaphragms was more significant in long span bridge.

In this paper, a new kind of brace system is proposed to strengthen the transversal connection of T beam bridges, which can be prefabricated in factory and installed in the field. Static experiments and finite element model shows that the newly proposed brace system can not only improve the load distribution between beams, but also enhance the integrity of T beam bridge.

* Corresponding author

IOP Conf. Series: Materials Science and Engineering **293** (2017) 012014 doi:10.1088/1757-899X/293/1/012014

2. Newly proposed brace system

The newly proposed brace system consists of two diagonal braces and one horizontal brace. They are connected by bolts with each other. The three braces connect with T beam bridge by a plate on the top and two brackets at the bottom. The brackets are fixed on the original T beam bridge by bolts. In order to study the strengthening effect of the proposed brace system, a T beam bridge was made by nylon and the brace system was made by aluminum alloy. Figure 1 shows the brace system and figure 2 is the nylon T beam bridge.

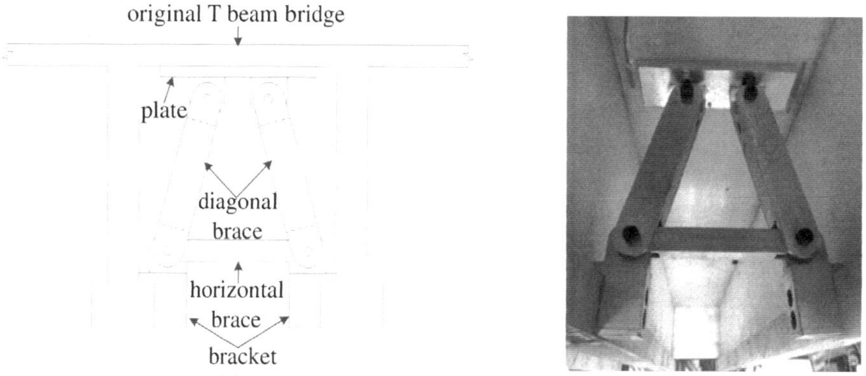

(a) (b)

Figure 1. Brace system.

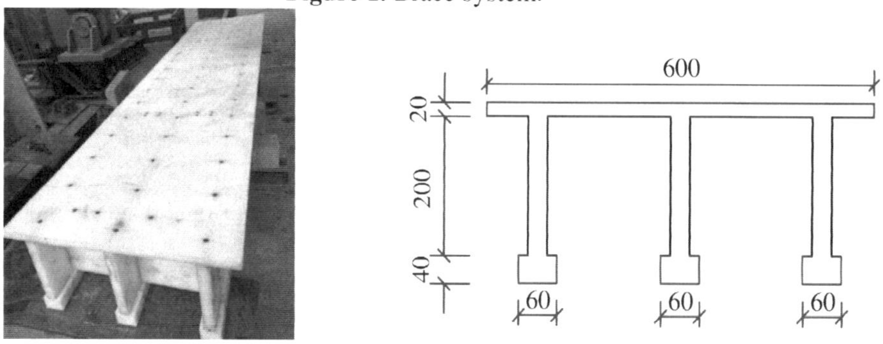

(a) (b)

Figure 2. Nylon T beam bridge (unit:mm).

3. Experiments

3.1. Material performance tests

As the T beam bridge was made of nylon, elastic modulus and Poisson's ratio of nylon were measured by tension tests. Elastic modulus was tested as figure 3 shows and figure 4 illustrates Poisson's ratio tests.

Figure 3. Elastic modulus tests.

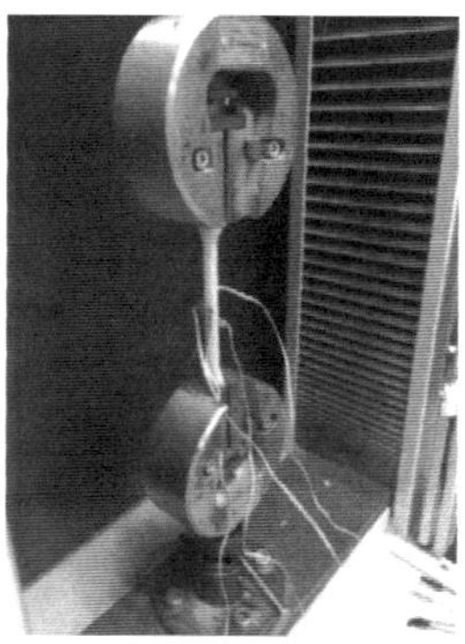

Figure 4. Poisson's ratio tests.

As shown in figure 5 below, the average of Poisson's ratio values is almost a constant around ν=0.386.

The average of elastic modulus is shown in figure 6 and is around E=1436Mpa. These values can be considered in numerical calculation.

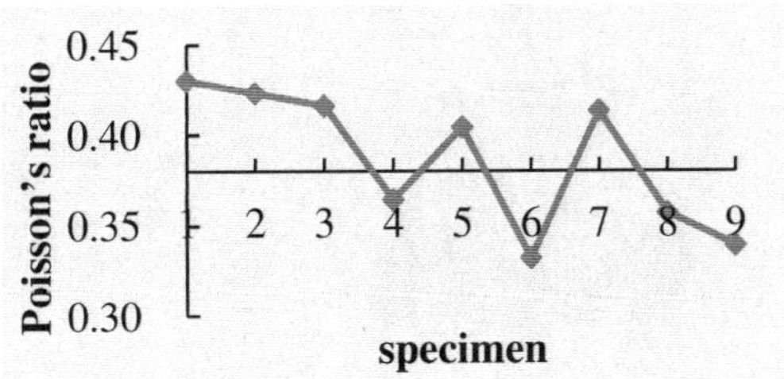

Figure 5. Poisson's ratio values.

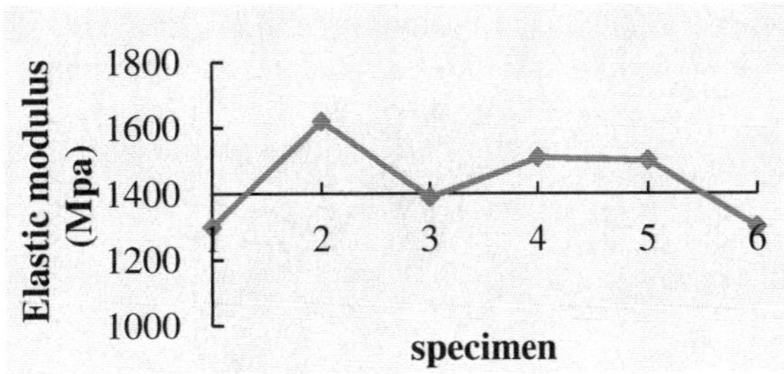

Figure 6. Elastic modulus values.

3.2. Static deflection tests

To measure the effect of the proposed brace system, static tests of the T beam bridge with and without brace system were carried out respectively. In the static tests, load was at mid span and on the top of the middle beam, which is shown in figure 7.

Figure 7. Load on bridge deck.

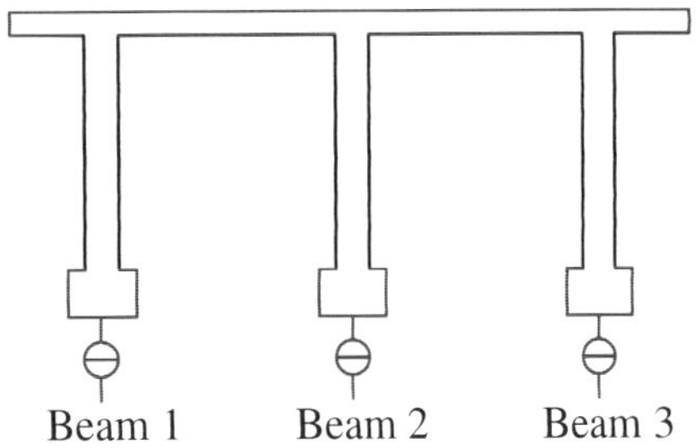

Figure 8. Mid span displacement meters locations.

Displacement meters were placed at the bottom of each beam to measure the mid span deflection, which is shown in figure 8. The mid span deflection of all beams before and after strengthening is shown in figure 9.

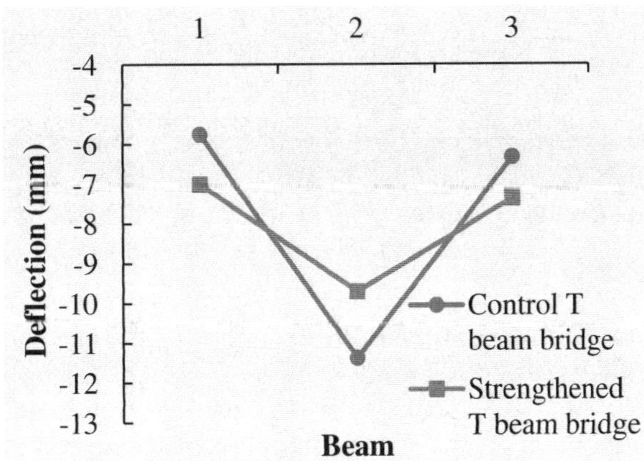

Figure 9. Experimental deflection.

At a same load, the maximum static deflection was decreased 14.9% when the T beam bridge was strengthened with the proposed brace system.

4. Finite element analysis

The ANSYS finite element software was used in this study to create models of the nylon T beam bridge and brace system to study the strengthening effect of the proposed brace system on transversal connection of T beam bridges.

The model's section is as given in figure 2(b) and the properties of nylon are as given in section 3.1. The elastic modulus and Poisson's ratio of aluminum alloy is 70000Mpa and 0.3[6], respectively.

Solid45 was adopted to represent the T beam bridge and brackets, whereas the diagonal braces and horizontal brace were represented by using beam4. Figure 10 shows the modeled control and strengthened T beam bridge in ANSYS software. The brackets were connected to T beam bridge at common node points and assumed a complete connection in finite element model.

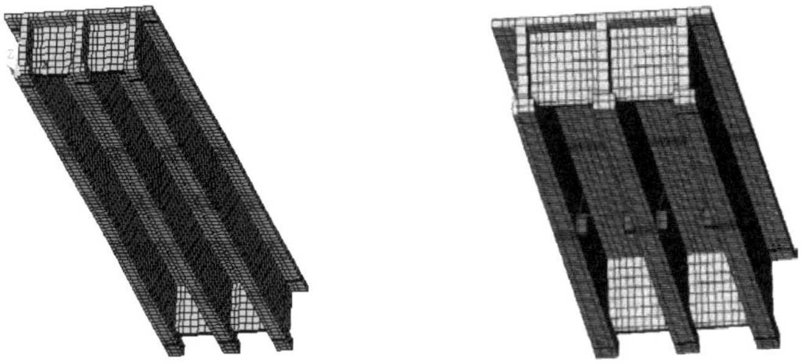

(a) (b)

Figure 10. Control and strengthened T beam bridge simulation detail.

Figure 11 shows the computed result of mid span deflection of three beams. At symmetrical loading, the maximum displacement decreased 19.1%. As the slippage of bolts in the experiments, the computed strengthening effect is better than that in experiments.

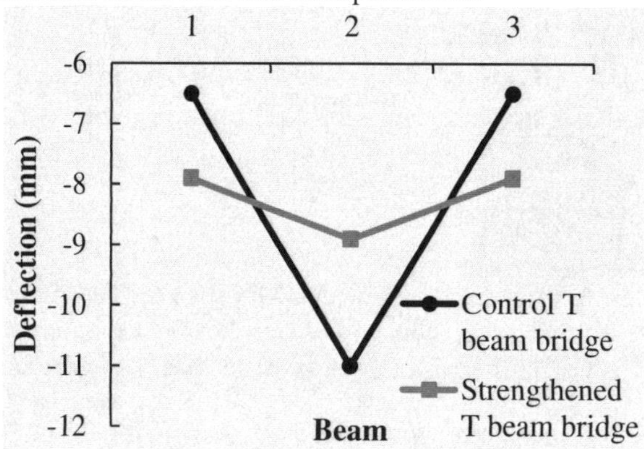

Figure 11. Computed deflection.

5. Conclusion

This paper presents a new strengthening method for transversal connection of T beam bridges. The proposed brace system can be manufactured in plant, avoiding interrupting the transportation. Experiments and finite element analysis showed that the proposed brace system can strengthen the connection between the adjacent beams and can solve the problem of "single slab bearing".

References

[1] Li H J, Zhao S C, Li W H, Yang Y and Zhang J Q 2014 Evaluation methods for transversal connection of precast beam bridges *Journal of China & Foreign Highway* **34** 124–8

[2] Xie Z Y, Wu Y B, Ma H W and Jia W W 2014 Research on transversal connection strengthening of Simply-supported T beams by transversal prestress *Highways & Automotive Applications* **3** 205–7

[3] Guo Z M 2009 Simply-supported T beams strengthening by bonded steel plates *Science and Technology of West China* **8** 14–5

[4] Niu Y W, Zhao Y, Rong S and Huang P M 2014 Effect of diaphragm on integrity of precast T-shaped girder bridge based on inner force analysis *Bridge maintenance, safety, management and life extension* 1193–7

[5] Li L, John Ma Z 2010 Effect of intermediate diaphragms on decked bulb-tee bridge system for accelerated construction *Journal of Bridge Engineering* **15** 715-722

[6] China Standard 2007 *Code for Design of Aluminum Structures* China p 17